THE SUPERSYMMETRIC DIRAC EQUATION

The Application to Hydrogenic Atoms

THE SUPERSYMMETRIC DIRAC EQUATION

The Application to Hydrogenic Atoms

Allen Hirshfeld

Technical University of Dortmund, Germany

Imperial College Press

Published by

Imperial College Press
57 Shelton Street
Covent Garden
London WC2H 9HE

Distributed by

World Scientific Publishing Co. Pte. Ltd.
5 Toh Tuck Link, Singapore 596224
USA office: 27 Warren Street, Suite 401-402, Hackensack, NJ 07601
UK office: 57 Shelton Street, Covent Garden, London WC2H 9HE

British Library Cataloguing-in-Publication Data
A catalogue record for this book is available from the British Library.

Paul Dirac photograph by A. Börtzells Tryckeri, courtesy of AIP Emilio Segre Visual Archives, E. Scott Bar and Weber Collections.

THE SUPERSYMMETRIC DIRAC EQUATION
The Application to Hydrogenic Atoms

ISBN-13 978-1-84816-797-1
ISBN-10 1-84816-797-0

Typeset by Stallion Press
Email: enquiries@stallionpress.com

Printed in Singapore.

I wish to dedicate this book to the memory of my father, Dr. Martin A. Hirshfeld, who taught me many things, among them the meaning of proof.

Preface

One morning, when I was an undergraduate, over a cup of coffee in the cafeteria of the Weizman Institute, where he was working on his PhD in physics, Charles Robinson opened my eyes to the wonders of the Dirac equation, and sowed the seed for this book.

I met Paul Dirac in the early 1960s, on the occasion of his visit to the Physics Department of the Tel-Aviv University. We talked about my physics project at that time (I was working on my Master's thesis), and I attended his talk, where he explained the difficulties in the foundations of quantum field theory. Afterwards he autographed his book for me.

I am very grateful to Jens Peder Dahl, Professor Emeritus of the Chemistry Department of the Technical University of Denmark, for the fruitful and encouraging correspondence on his work.

I wish to thank Dirk Fischer for his assistance with the technical aspects of the preparation of the manuscript, and especially for his help in the preparation of the figures.

Contents

List of Figures

Chapter 1
Introduction

> *"He succeeded in combining quantum mechanics and the theory of relativity,... and introduced the idea of an antiparticle... Unfortunately, only the expert can appreciate Dirac's awesome work. To this day, that work has not lost any of its splendor."*

Martinus Veltman, *Facts and Mysteries in Elementary Particle Physics*, World Scientific Publishing, Singapore, 2003.

The sentiment expressed in this quotation is certainly true. But let us pause to consider the phrase "only the expert can appreciate Dirac's awesome work . . . ". We expect a student beginning to study quantum field theory to understand the working of relativistic quantum theory as it applies to free particles. The next interesting case, that of a particle in a Coulomb field, is given at best a somewhat cursory treatment in most texts. But it was precisely this problem that was the test of the new theory — could it explain the fine structure in the spectrum of hydrogen? The fact of this success and the detailed structure that it predicts were the hallmark of the theory for many years. It was the work devoted to testing this structure, with the highest possible precision, which revealed the need for the radiative corrections of quantum electrodynamics and provided us with the most accurate evaluation of the fine structure, which we possess to the present day, and one of the most impressive successes in all of physics.

It is thus highly desirable that the modern student of relativistic quantum mechanics should have an appreciation of this decisive episode in the physics of the twentieth century. The advances of the theory in recent years have led to certain simplifications with respect to Dirac's time. In particular, the discovery that supersymmetry plays a central role in the structure of the theory of a particle in a Coulomb field means that while gaining a mastery of this topic, students are at the same time gaining experience that will be of importance in appreciating the questions with which their future work will be concerned, namely, does supersymmetry play a role in the structure of the universe? Even if this question is, contrary to expectation, answered in the negative, the knowledge of supersymmetry is valuable because it allows a better understanding of many solvable problems in physics, and because of the role it plays in mathematics.

Thus, the purpose of the present book is to impart to the reader the necessary "expertise" for appreciating the application of the Dirac theory to the hydrogen atom. It will become clear that most of the concepts can be introduced in the more familiar, non-relativistic context in an elementary fashion. We will also become acquainted with a previous attempt to derive the formula for the fine structure of hydrogen, due to Sommerfeld. In this work, the *identical* formula for the energy levels of the hydrogen atom was obtained, 12 years *before* Dirac. This was before the Schrödinger wave equation for the non-relativistic hydrogen atom, and before the discovery of the spin of the electron! It is built on the Bohr model of the atom, but takes into account the relativistic motion of the electron. We shall see that it characterizes the orbits by their *eccentricity*, in a way that agrees with the Dirac theory.

We also take advantage, in our exposition, of the opportunity provided to relive the highlights of the history of physics. Kepler's observation that Mars, in its motion around the Sun, follows an elliptical orbit, marks the beginning of physics as a quantitative science. In subsequent years, this system of two bodies interacting through the influence of a central force has served as a model of such disparate motions as the collision of galaxies and the orbits of electrons about the nuclei of atoms. It is the latter development that concerns us here.

For a particle moving in a central force field the angular momentum vector $\mathbf{L}$ is conserved. If the force is inversely proportional to the distance squared there is another conserved vector, the Laplace vector $\mathcal{A}$. The Laplace vector lies along the fixed axis of the orbital motion, and its magnitude is a measure of the eccentricity $e = |\mathcal{A}|$, which characterizes the deviation of the orbit from a circle. The Laplace vector turns out to be one of the central concepts of the book. The three components of the angular momentum, complemented by the three components of the Laplace vector, form, properly normalized, the generators of an SO(4) symmetry group.

The Rutherford [Rut11] picture of the atom as a miniature solar system suggested an approach to understanding the spectral lines of light emitted in atomic transitions. With Bohr's [Boh13] "quantum rules", based on the postulate of a smallest unit of orbital angular momentum, the era of the "old quantum theory" was initiated.

A more refined version of the Bohr model is a non-relativistic form of Sommerfeld's model [Som16], based on the Sommerfeld–Wilson quantization rules. The set of elliptical orbits predicted in the classical theory is reproduced, along with the formula for their eccentricity.

The beginning of the "new quantum theory" was marked, as a matter of fact, by Pauli's [Pau26] group theoretical treatment of the hydrogen atom. Pauli's analysis is based on the existence of the Laplace vector $\mathcal{A}$ as a conserved quantity of the motion of a point particle moving under the influence of an inverse-quadratic force. Its three components form, together with the three components of the orbital angular momentum, the generators of the Lie algebra so(4). The quadratic Casimir operator of this algebra then yields a formula for the spectrum of the bound-state energy levels of hydrogenic atoms. The advantage of this approach over Schrödinger's [Sch26] method of solving a differential equation is that it automatically yields an understanding of the degeneracy of the spectral lines, which is treated in the Schrödinger approach as an "accidental" symmetry.

A full analysis of the problem of an electron with spin in a Coulomb field involves the two-component Pauli formalism. It is only necessary to replace $\mathbf{p}$ by $(\boldsymbol{\sigma} \cdot \mathbf{p})$ to get the correct gyromagnetic ratio. The inclusion of the spin variables provides a formalism in which quadratic operators of the theory factorize. The effectiveness of such factorizations was already noted by Hull and Infeld [HI51], whose work was in a sense a precursor of the theory of supersymmetry, which arose originally in the quantum field theory of elementary particles. The Schrödinger equation for the radial component of the wave function is seen to be the product of two linear differential operators, whose eigenvalues are related to the magnitude of the operator $A_{nr} = (\boldsymbol{\sigma} \cdot \mathcal{A})$, in a way that is the quantum version of the eccentricity formula $e = |\mathcal{A}|$. Couching the results of the treatment of spin in a supersymmetric language leads to an understanding of the full degeneracy of the hydrogen spectrum as a consequence of the SO(4) × S(2) supersymmetry group.

When we ask for an understanding of the *fine structure* of the spectrum we move on to the treatment of the relativistic spectrum. Today we undertake a study of the solutions of the Dirac equation describing an electron in a Coulomb potential. In this book we first follow in the footsteps of Sommerfeld, who undertook a treatment of the relativistic problem, although he had at his disposal neither the modern quantum theory nor the notion of electron spin. It is all the more astonishing that he obtained a formula for the spectrum of hydrogen that agrees with the results of the Dirac theory. The Sommerfeld theory not only gives the correct formula for the energy of the hydrogen orbits, it also describes their relativistic eccentricity. One of the purposes of this book is to explain this success, based on the work of Biedenharn [Bie83].

When we go over to the relativistic problem the SO(4) symmetry breaks down, but a remnant of the Laplace vector survives, namely the Johnson–Lippmann operator A [JL50]. The Johnson–Lippmann operator is, in the non-relativistic approximation, the operator A_{nr}. In this way the degeneracy of the spectrum reduces to that of SO(3) × S(2), with SO(3) the symmetry generated by the total angular momentum $\mathbf{J} = \mathbf{L} + \mathbf{S}$, where $\mathbf{S}$ is the spin vector that generates the SO(3) rotation group, and S(2) is the supersymmetry group.

By describing the Dirac theory in a supersymmetric framework we are able to derive the solutions directly, without resorting to the heuristic methods employed formerly. Here we follow the work of Dahl and Jørgensen [DJ95]. The solutions of the Dirac equation are obtained by investigating two different extensions of the solution space. We describe the relation of the solutions to the second-order Kramer's equation, which was used formerly to construct solutions of the Dirac equation.

Finally, the non-relativistic approximations of the solutions found for the Dirac equation are shown to reduce to the solutions of the Pauli equation. This justifies the choice of sign for the solutions, and allows a check on the various normalization factors used in this book.

The Appendices deal with some of the features of the confluent hypergeometric functions, and their relation to the wave functions of the Pauli and Dirac equations.

The SO(3) × S(2) symmetry of the Dirac theory is broken by the radiative corrections of quantum field theory, notably in the Lamb shift [LR47], which describes the energy difference between the $2s_{\frac{1}{2}}$ and $2p_{\frac{1}{2}}$ levels of hydrogen. Because of this, the Dirac theory of bound states belongs in a course of study after non-relativistic quantum mechanics and before the introduction to the quantum theory of interacting fields. It is for such students that the present book is intended.

A note on units

This book uses the system of units in which $\hbar$ and the velocity of light are set to unity: $\hbar = c = 1$, where $\hbar = h/(2\pi)$, and h is Planck's constant. The Einstein convention is used throughout, according to which an index that appears in an expression twice is summed over. The fine structure constant is $\alpha = e^2/4\pi$, where e is the charge of the electron.

Chapter 2
The Classical Kepler Problem

"It is not eighteen months since I first caught a glimpse of the light, three months since the dawn, very few days since the unveiled Sun, most admirable to gaze upon, burst upon me. Nothing restrains me; I shall indulge my sacred fury; I shall triumph over mankind by the honest confession that I have stolen the golden vases of the Egyptians to build up a tabernacle for my God far from the confines of Egypt. If you forgive me, I rejoice; if you are angry, I can bear it; the die is cast, the book is written, to be read either now or by posterity, I care not which; it may well wait a century for a reader, as God himself has waited six thousand years for someone to behold his work."

Johannes Kepler, *Gesammelte Werke*, 18 volumes, C. H. Beck, Munich, 1937–1969.

We are all familiar with the modern understanding of our solar system. The Sun sits at the center of a ring of essentially concentric orbits, each planet following an elliptical path around the Sun, which sits not at the center but at one focus of the ellipse. It is the picture shown in Figure 2.1. No one has ever seen this picture. Even today there are no satellites sitting at that vantage point tracking the orbits and plotting their paths, much less in the seventeenth century when this picture was first proposed by Johannes Kepler. Kepler and the astronomers before and since him have constructed their understanding of the solar system by watching the Sun and planets, observable as points of light, move across the sky. We only see this two-dimensional projection of their paths.

Kepler's "unveiled Sun" was the realization that the observational data of Tycho Brahe, his mentor, could be explained by the following three laws:

- A planetary orbit is an ellipse with the Sun at one focus.
- A planetary orbit sweeps out equal areas in equal times.
- The square of the period of the orbit is directly proportional to the cube of its mean distance.

Kepler's laws of motion were explained by Newton. Newton was able to show that they could be deduced from his universal law of gravitation, involving a central

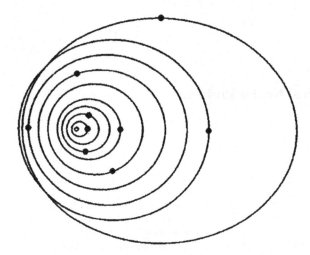

Figure 2.1. The solar system as we understand it.

force that falls off as the square of the distance. Taking the mass of the central attractor as infinite and the planet as a point particle leads to the problem of a point mass in a central potential field. The motion takes place in a fixed plane because of the conservation of angular momentum. The fact that the major axis of the ellipse remains stationary for a $1/r$ potential was known, by the time of Laplace, to be due to the conservation of the Laplace (Runge–Lenz) vector [Gol75], whose magnitude is equal to the eccentricity of the ellipse. With the model of an atom as a miniature solar system these concepts became relevant to the motion of electrons about the nucleus.

2.1 Central Forces

The velocity of the point P in space is

$$v = \frac{dr}{dt} = \frac{dr}{dt}\hat{r} + r\frac{d\hat{r}}{dt} = \frac{dr}{dt}\hat{r} + r\frac{d\theta}{dt}\hat{\theta}, \qquad (2.1)$$

where $r = r\hat{r}$ is the position vector and $\hat{r}$ is a unit vector on the direction of r. $\hat{\theta}$ is a unit vector perpendicular to $\hat{r}$:

$$\hat{r} = (\cos\theta(t), \sin\theta(t)), \quad \hat{\theta} = (-\sin\theta(t), \cos\theta(t)). \qquad (2.2)$$

The acceleration is

$$\boldsymbol{a} = \frac{d\boldsymbol{v}}{dt} = \frac{d^2r}{dt^2}\hat{\boldsymbol{r}} + 2\frac{dr}{dt}\frac{d\theta}{dt}\hat{\boldsymbol{\theta}} + r\frac{d^2\theta}{dt^2}\hat{\boldsymbol{\theta}} + r\frac{d\theta}{dt}\left(-\frac{d\theta}{dt}\hat{\boldsymbol{r}}\right)$$

$$= \left(\frac{d^2r}{dt^2} - r\left(\frac{d\theta}{dt}\right)^2\right)\hat{\boldsymbol{r}} + \frac{1}{r}\frac{d}{dt}\left(r^2\frac{d\theta}{dt}\right)\hat{\boldsymbol{\theta}}. \qquad (2.3)$$

Equation (2.3) tells us that the acceleration is entirely radial, that is, parallel to $\boldsymbol{r}$, if and only if

$$\frac{d}{dt}\left(r^2\frac{d\theta}{dt}\right) = 0, \qquad (2.4)$$

which is equivalent to saying that $r(t)^2 d\theta/dt$ is a constant independent of t.

Combining this with the next lemma gives a proof of Kepler's second law:

Lemma. If the position of a particle over time is described by the vector function $r(t)$, then the rate at which the radial vector sweeps out area is given by

$$\frac{dA}{dt} = \frac{r(t)^2}{2}\frac{d\theta}{dt}. \qquad (2.5)$$

Proof. Given a circle with center at the origin and radius r, the area swept out by the radius as it moves through an angle of $\Delta\theta$ is given by $(r^2/2)\Delta\theta$. It follows that if ΔA is the area swept out by the radial vector from time s to time t, and if the distance from the origin stays constant during this time interval, then

$$\Delta A = \frac{r^2}{2}\Delta\theta, \qquad (2.6)$$

where $\Delta\theta = \theta(t) - \theta(s)$. If r does not stay constant, then we can find two points in the interval $[s, t]$, call them t_1 and t_2, where r takes on its minimum and maximum values, respectively, over this interval: $r(t_1) \leq r \leq r(t_2)$. It follows that

$$\frac{r(t_1)^2}{2}\Delta\theta \leq \Delta A \leq \frac{r(t_2)^2}{2}\Delta\theta. \qquad (2.7)$$

We now divide by $\Delta t = t - s$:

$$\frac{r(t_1)^2}{2}\frac{\Delta\theta}{\Delta t} \leq \frac{\Delta A}{\Delta t} \leq \frac{r(t_2)^2}{2}\frac{\Delta\theta}{\Delta t}, \qquad (2.8)$$

and take the limit as s approaches t. This forces t_1 and t_2 to also approach t and yields

$$\frac{r(t)^2}{2}\frac{d\theta}{dt} \leq \frac{dA}{dt} \leq \frac{r(t)^2}{2}\frac{d\theta}{dt}. \qquad (2.9)$$

Lemma. Let $\mathbf{r}(t)$ be the position of a particle at time t, and $\mathbf{v}(t)$ its velocity. The *momentum* of the particle is $\mathbf{p} = m\mathbf{v}$, where m is its mass. If the particle is moving under the influence of a central force:

$$\boldsymbol{F} = \frac{d\mathbf{p}}{dt} = f(r)\hat{\boldsymbol{r}}, \qquad (2.10)$$

where f is an arbitrary function of r, then the orbital angular momentum

$$\mathbf{L} = \mathbf{r} \times \mathbf{p} \qquad (2.11)$$

is a constant vector of magnitude

$$L = |\mathbf{L}| = mr(t)^2 \frac{d\theta}{dt}. \qquad (2.12)$$

Proof. The fact that $\mathbf{L}$ is a constant vector is immediate:

$$\frac{d\mathbf{L}}{dt} = \frac{d}{dt}(\mathbf{r} \times \mathbf{p}) = m\mathbf{v} \times \mathbf{v} + f(r)\mathbf{r} \times \hat{\boldsymbol{r}} = \mathbf{0}. \qquad (2.13)$$

To prove Eq. (2.12) we write

$$\mathbf{L} = mr\hat{\boldsymbol{r}} \times \left(\frac{dr}{dt}\hat{\boldsymbol{r}} + r\frac{d\theta}{dt}\hat{\boldsymbol{\theta}} \right) = mr^2 \frac{d\theta}{dt}(\hat{\boldsymbol{r}} \times \hat{\boldsymbol{\theta}}), \qquad (2.14)$$

so that

$$L = mr^2 \frac{d\theta}{dt}. \qquad (2.15)$$

Since

$$\mathbf{L} \cdot \mathbf{r} = (\mathbf{r} \times \mathbf{p}) \cdot \mathbf{r} = (\mathbf{r} \times \mathbf{r}) \cdot \mathbf{p} = 0, \qquad (2.16)$$

the motion is in a plane perpendicular to $\mathbf{L}$.

2.2 The Laplace Vector

We now consider the case of a force inversely proportional to the square of the distance from the origin:

$$\boldsymbol{F} = \frac{d\mathbf{p}}{dt} = -\left(\frac{k}{r^2} \right) \hat{\boldsymbol{r}}. \qquad (2.17)$$

Kepler's first law is a consequence of the following theorem:

Theorem. Let $\mathbf{r}(t)$ denote the position of a particle moving under the influence of a force inversely proportional to the square of the distance from the origin. Then there exists a constant vector $\mathcal{A}$, the Laplace vector, such that

$$|\mathbf{r}| + \mathbf{r} \cdot \mathcal{A} = \frac{L^2}{mk}. \tag{2.18}$$

Equivalently, if (r, θ) is the position in polar coordinates, then

$$r(1 + e\cos\theta) = \frac{L^2}{mk}, \tag{2.19}$$

where the eccentricity e is equal to the magnitude of $\mathcal{A}$: $e = |\mathcal{A}|$. We recognize Eq. (2.19) as the equation of a conic section: an ellipse, parabola, or hyperbola. In particular, if $e < 1$, then it is the equation of an ellipse with one focus at the origin.

Proof. We note

$$L^2 = \mathbf{L} \cdot (\mathbf{r} \times \mathbf{p}) = (\mathbf{p} \times \mathbf{L}) \cdot \mathbf{r}. \tag{2.20}$$

Now

$$\frac{d}{dt}(\mathbf{p} \times \mathbf{L}) = \frac{d\mathbf{p}}{dt} \times \mathbf{L} = \left(-\frac{k}{r^2}\hat{r}\right) \times \left(mr^2\frac{d\theta}{dt}\hat{r} \times \hat{\theta}\right)$$

$$= -mk\frac{d\theta}{dt}[\hat{r} \times (\hat{r} \times \hat{\theta})] = mk\frac{d\theta}{dt}\hat{\theta} = \frac{d}{dt}(mk\hat{r}). \tag{2.21}$$

This means that the time derivative of $\mathbf{p} \times \mathbf{L} - mk\hat{r}$ is zero, so

$$\mathcal{A} = \frac{1}{mk}(\mathbf{p} \times \mathbf{L}) - \hat{r} \tag{2.22}$$

is a constant vector, and

$$\mathbf{p} \times \mathbf{L} = mk(\hat{r} + \mathcal{A}). \tag{2.23}$$

Combining this result with Eq. (2.20), we see that

$$L^2 = (\mathbf{p} \times \mathbf{L}) \cdot \mathbf{r} = mk(|\mathbf{r}| + \mathbf{r} \cdot \mathcal{A}), \tag{2.24}$$

or

$$|\mathbf{r}| + \mathbf{r} \cdot \mathcal{A} = \frac{L^2}{mk}. \tag{2.25}$$

Equation (2.19) follows from the equalities $|\mathbf{r}| = r$ and $\mathbf{r} \cdot \mathcal{A} = re\cos\theta$.

An ellipse with foci F_1 and F_2 can be defined as the locus of points P for which $|F_1 P| + |F_2 P| = 2a$, with a the semimajor axis. The point P is described by the vector function $\boldsymbol{r}(t) = r(t)\hat{r}(t)$, with $r(t) = \sqrt{x(t)^2 + y(t)^2}$ and $\hat{r}(t) = (\cos\theta(t), \sin\theta(t))$. See Figure 2.2.

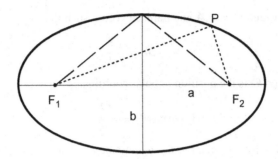

Figure 2.2.   An ellipse.

Lemma. The general equation of an ellipse with one focus at the origin and semimajor axis along the x-axis is given in polar coordinates by

$$r(1 + e\cos\theta) = c, \tag{2.26}$$

where e and c are constants, $0 < e < 1$, $c > 0$. The semimajor axis is

$$a = \frac{c}{1 - e^2}, \tag{2.27}$$

and the semiminor axis is

$$b = \frac{c}{\sqrt{1 - e^2}}. \tag{2.28}$$

The *eccentricity* of the ellipse is

$$e = \sqrt{1 - (b/a)^2}. \tag{2.29}$$

Proof. Equation (2.26) can be rewritten as

$$r = c - er\cos\theta. \tag{2.30}$$

We convert this to Cartesian coordinates ($x = r\cos\theta$, $y = r\sin\theta$) and square each side to obtain

$$x^2 + y^2 = c^2 - 2ecx + e^2 x^2, \tag{2.31}$$

or

$$(1 - e^2)x^2 + 2ecx + y^2 = c^2. \tag{2.32}$$

Now add $c^2 e^2/(1 - e^2)$ to each side:

$$(1 - e^2)x^2 + 2ecx + \frac{c^2 e^2}{1 - e^2} + y^2 = c^2 + \frac{c^2 e^2}{1 - e^2}, \tag{2.33}$$

or

$$(1 - e^2)(x + ea)^2 + y^2 = \frac{c^2}{1 - e^2}, \qquad (2.34)$$

which may be rewritten as

$$\frac{(x + ea)^2}{a^2} + \frac{y^2}{b^2} = 1. \qquad (2.35)$$

This is the equation of an ellipse in Cartesian coordinates. The center of the ellipse is at $(-ae, 0)$, which is a distance ae from the origin. The foci are bisected by a line that intersects the ellipse at two points. At these points the distance from the foci is $\sqrt{(ae)^2 + b^2} = a$, and the sum of the distances from the foci, which is the same for all points of the ellipse, is $2a$.

The *perigee*, which is the point of closest approach, corresponds to $\theta = 0$, so

$$r_{\min} = a(1 - e) = \frac{L^2}{mk(1 + e)}. \qquad (2.36)$$

The *apogee*, which is the farthest distance, corresponds to $\theta = \pi$, so

$$r_{\max} = a(1 + e) = \frac{L^2}{mk(1 - e)}. \qquad (2.37)$$

The *mean distance* is $\frac{1}{2}(r_{\min} + r_{\max}) = a$.

The Laplace vector $\mathcal{A}$ is perpendicular to the angular momentum,

$$\mathcal{A} \cdot \mathbf{L} = 0, \qquad (2.38)$$

so $\mathcal{A}$ lies in the plane of motion of **r**. As a matter of fact, $\mathcal{A}$ lies along the line joining the two foci.

We express the eccentricity in terms of the constants of the motion, the energy E, and the angular momentum L. We use the identities

$$(a) \quad (\boldsymbol{p} \times \boldsymbol{L}) \cdot (\boldsymbol{p} \times \boldsymbol{L}) = p^2 L^2,$$

$$(b) \quad (\boldsymbol{p} \times \boldsymbol{L}) \cdot \boldsymbol{r} = \boldsymbol{L} \cdot (\boldsymbol{r} \times \boldsymbol{p}) = L^2. \qquad (2.39)$$

We then have

$$|\mathcal{A}|^2 = \frac{p^2 L^2}{m^2 k^2} - \frac{2L^2}{mkr} + 1 = \frac{2L^2}{mk^2}\left(\frac{p^2}{2m} - \frac{k}{r}\right) + 1 = \frac{2L^2 E}{mk^2} + 1 = e^2, \qquad (2.40)$$

where

$$E = \frac{p^2}{2m} - \frac{k}{r}$$ (2.41)

is the constant energy of a particle in a fixed orbit.

From the equation for the perigee we have

$$r_{\min} = a(1 - e) = \frac{L^2}{mk(1 + e)},$$ (2.42)

or

$$1 - e^2 = \frac{L^2}{mka}.$$ (2.43)

· Comparing this to Eq. (2.40):

$$1 - e^2 = -\frac{2L^2 E}{mk^2},$$ (2.44)

we get

$$E = -\frac{k}{2a}.$$ (2.45)

The energy of a particle in orbit depends only on the semimajor axis of the ellipse.

We can now derive Kepler's third law. By Kepler's first law a particle in orbit moves on an ellipse. The semiminor axis of the ellipse is $b = a\sqrt{1 - e^2}$. By Kepler's second law the area of the ellipse is

$$A = \int_0^\tau \frac{dA}{dt}\, dt = \frac{L}{2m}\, \tau,$$ (2.46)

where τ is the period of a complete revolution. Hence

$$A = \frac{L}{2m}\tau = \pi ab = \pi a^2 \sqrt{1 - e^2} = \pi a^2 \frac{L}{k}\sqrt{\frac{-2E}{m}} = \pi a^{3/2}\frac{L}{\sqrt{km}},$$ (2.47)

or

$$\tau = 2\pi a^{3/2}\sqrt{\frac{m}{k}}.$$ (2.48)

The square of the period of the orbit is therefore proportional to the cube of a, which is the mean distance.

Notes on Chapter 2

The conservation of the Laplace vector and the interpretation of its length as eccentricity are first mentioned by Jacob Hermann, in a letter to Bernoulli. It was then used by Laplace in his treatise on mechanics. It was rediscovered by Hamilton in the nineteenth century, and included in textbooks by Gibbs and Runge. Pauli learned about it from his teacher Lenz, and after Pauli's quantum mechanical treatment of the hydrogen atom (Chapter 8) it is commonly referred to as the Runge–Lenz vector. The early history of the Laplace vector was traced by Goldstein [Gol75] [Gol76]. See also the Wikipedia page at http://en.wikipedia.org/wiki/Laplace–Runge–Lenz_vector. The presentation here follows Bressoud [Bre91]. The origin of the Laplace vector as the boost-generator of the relativistic extension of the theory has been proposed by Dahl [Dah95].

Chapter 3
Symmetry of the Classical Problem

"Symmetry, as wide or as narrow as you may wish to define it, is one idea by which man through the ages has tried to comprehend and create order, beauty, and perfection."

Hermann Weyl, in *Symmetry*, Princeton University Press, Princeton, 1952.

In this chapter we make an excursion into mathematics. We first review the definitions of Lie groups and Lie algebras, because these are the mathematical basis of symmetry considerations in physics. We show how these concepts fit into the Hamiltonian formulation of classical mechanics. We then return to the problem of a central force which is inversely proportional to the square of the distance, and show that the conservation of the Laplace vector, together with the conservation of angular momentum, is equivalent to the existence of an SO(4) symmetry group that acts on functions of the phase space.

3.1 Lie Groups and Lie Algebras

Let A be a linear operator on a finite dimensional vector space V. If $V = \mathbb{R}^n$ then A is an $n \times n$ matrix. The operator valued function $t \mapsto e^{tA}$ is defined by

$$e^{tA} = I + tA + \frac{1}{2}t^2 A^2 + \frac{1}{3!}t^3 A^3 + \cdots \tag{3.1}$$

which converges for all t. For a fixed A, the function e^{tA} behaves like the ordinary exponential function:

$$e^{tA}|_{t=0} = I, \quad e^{(t+s)A} = e^{tA}e^{sA}, \quad \frac{d}{dt}e^{tA} = Ae^{tA}. \tag{3.2}$$

e^{tA} is a curve of linear operators, and the LHS of the last equation gives the tangent vector to this curve at time t. At $t = 0$ we have

$$\frac{d}{dt}e^{tA}|_{t=0} = A. \tag{3.3}$$

If R is some other invertible operator then

$$(RAR^{-1})^n = RA^n R^{-1} \tag{3.4}$$

for all n, so

$$e^{t(RAR^{-1})} = Re^{tA}R^{-1}. \tag{3.5}$$

In particular, the tangent vector to $Re^{tA}R^{-1}$, which goes through the identity I at $t = 0$, is RAR^{-1}.

Suppose A is fixed and $R(s)$ is a curve of matrices that passes through I at $s = 0$ and is differentiable in s. Then $R(s)$ is invertible for s close to the identity, and differentiating

$$R(s)R(s)^{-1} = I \tag{3.6}$$

at $s = 0$ gives

$$\frac{d}{ds}R^{-1}(s)|_{s=0} = -\frac{d}{ds}R(s)|_{s=0} = -B, \tag{3.7}$$

and

$$\frac{d}{ds}[R(s)AR(s)^{-1}]_{s=0} = BA - AB, \tag{3.8}$$

by Leibniz's rule. Define the *Lie bracket* $[\,,\,]$ by

$$[B, A] = BA - AB. \tag{3.9}$$

The Lie bracket is bilinear, antisymmetric, and satisfies Jacobi's identity:

$$\begin{aligned}
[A, [B, C]] &= [A, BC - CB] = [A, BC] - [A, CB] \\
&= [A, B]C + B[A, C] - [A, C]B - C[A, B] \\
&= (AB - BA)C + B(AC - CA) \\
&\quad - (AC - CA)B - C(AB - BA) \\
&= [[A, B], C] + [B, [A, C]].
\end{aligned} \tag{3.10}$$

A subgroup G of the group of all invertible matrices is said to be a linear *Lie group* if it is also a closed manifold.

For example, let O(n) be the $n \times n$ matrices that satisfy $OO^T = I$, where O^T denotes the transposed matrix. This defines O(n) as a subgroup of the n^2 dimensional space of $(n \times n)$-matrices. Let $R(s)$ be a curve of orthogonal matrices with $R(0) = I$ and $R'(0) = B$. Differentiate $R(s)R(s)^T = I$ at $s = 0$, and obtain

$$B + B^T = 0, \tag{3.11}$$

so B is antisymmetric. The dimension of the tangent space, and hence the manifold $O(n)$, is thus $n(n-1)/2$. $O(3)$ is a three-dimensional subgroup of the nine-dimensional group of all invertible 3×3 matrices. $O(4)$ is six-dimensional.

Conversely, let A be an antisymmetric matrix.

$$\left(e^{tA}\right)^T = e^{tA^T}, \tag{3.12}$$

since $(A^T)^n = (A^n)^T$.

$$\frac{d}{dt}\left[e^{tA}(e^{tA})^T\right] = \frac{d}{dt}[e^{tA}e^{tA^T}] = e^{tA}(A + A^T)e^{tA^T}. \tag{3.13}$$

But $A + A^T = 0$. Hence

$$\frac{d}{dt}[e^{tA}(e^{tA})^T] = 0. \tag{3.14}$$

At $t = 0$, $e^{tA}(e^{tA})^T|_{t=0} = I$, hence for all t

$$e^{tA}(e^{tA})^T = I. \tag{3.15}$$

If A is antisymmetric, e^{tA} is orthogonal.

Let G denote the Lie group $O(n)$ and $\mathcal{G}$ the space of antisymmetric matrices. Then $\mathcal{G}$ is the tangent space of G at the identity. So if $R(s)$ is a differentiable curve of matrices in G and $R(0) = I$, then $R'(0) \in \mathcal{G}$.

If $A \in \mathcal{G}$ then the curve e^{tA} is a *one-parameter subgroup* of G.

The *Lie bracket* of two antisymmetric matrices is antisymmetric:

$$[A, B]^T = (AB - BA)^T = B^T A^T - A^T B^T$$
$$= [B^T, A^T] = -[A^T, B^T] = -[A, B]. \tag{3.16}$$

But it is more instructive to derive this from the statements above. Indeed, let $B \in \mathcal{G}$. Then $R(s) = e^{sB}$ is a curve of matrices in G. Hence for each fixed s, conjugation by $R(s)$ carries G into itself, and carries I into itself, and hence must carry the tangent space $\mathcal{G}$ into itself. Hence if $A \in \mathcal{G}$, the curve $C(s) = R(s)AR(s)^{-1}$ lies in G; $C(s) \in G$ for all s. But then $C(s+h) - C(s) \in G$ for any s and h. Dividing by h and passing to the limit implies $C'(s) \in \mathcal{G}$. At $s = 0$ this tells us that for $B \in \mathcal{G}$ and $A \in \mathcal{G}$ also $[B, A] \in \mathcal{G}$.

The corresponding facts are true in general: If G is a linear Lie group then its tangent space $\mathcal{G}$ at I is the space spanned by $A \in \mathcal{G}$, and the curve e^{tA} is a one-parameter subgroup of G. We say that $\mathcal{G}$ is the *Lie subalgebra* of the space of all matrices.

More generally, a linear space $\mathcal{G}$ with a bracket mapping $\mathcal{G} \times \mathcal{G} \mapsto \mathcal{G}$ is called a *Lie algebra* if the bracket is bilinear, antisymmetric, and satisfies the Jacobi identity. A group G that is a finite-dimensional manifold, where the group multiplication is consistent with the manifold structure in an obvious way, is called a *Lie group*. The tangent space to G at I (in the sense of differentiable manifolds) then inherits the structure of a Lie algebra. Given any finite-dimensional Lie algebra $\mathcal{G}$ over the real numbers, there always exists a Lie group G having $\mathcal{G}$ as its Lie algebra. All such G will be "locally isomorphic", but may differ in their global structure. There is a unique G that is connected and simply connected having $\mathcal{G}$ as its Lie algebra.

3.2 Some Special Lie Algebras

We now discuss in more detail the Lie algebras with which we shall be concerned.

$o(3)$

Let δ_i denote infinitesimal rotations about the x_i-axis, $i = 1, 2, 3$, so that

$$\delta_1 = \begin{pmatrix} 0 & 0 & 0 \\ 0 & 0 & -1 \\ 0 & 1 & 0 \end{pmatrix}, \quad \delta_2 = \begin{pmatrix} 0 & 0 & 1 \\ 0 & 0 & 0 \\ -1 & 0 & 0 \end{pmatrix}, \quad \delta_3 = \begin{pmatrix} 0 & -1 & 0 \\ 1 & 0 & 0 \\ 0 & 0 & 0 \end{pmatrix}. \quad (3.17)$$

Then

$$[\delta_1, \delta_2] = \delta_3, \quad [\delta_2, \delta_3] = \delta_1, \quad [\delta_3, \delta_1] = \delta_2. \quad (3.18)$$

This can be expressed more succinctly by $[\delta_i, \delta_j] = \epsilon_{ijk}\delta_k$, where

$$\epsilon_{ijk} = \begin{cases} 1 & \text{for } (i, j, k) \text{ an even permutation of } (1, 2, 3) \\ -1 & \text{for } (i, j, k) \text{ an odd permutation of } (1, 2, 3) \\ 0 & \text{otherwise.} \end{cases} \quad (3.19)$$

The δ_i are clearly linearly independent and so form a basis of o(3). So if we write the most general element of o(3) as $\mathbf{a} = a_1\delta_i + a_2\delta_2 + a_3\delta_3$ then the preceding equations for the brackets of the basis elements show that the Lie product in o(3) can be identified with the usual vector product,

$$[\mathbf{a}, \mathbf{b}] = \mathbf{a} \times \mathbf{b}. \tag{3.20}$$

Notice that this operation can also be described by thinking of **a** as a matrix and **b** as a vector and multiplying the vector **b** by the matrix **a**. For example, the fact that $[\delta_1, \delta_2] = \delta_3$ can be expressed as

$$\begin{pmatrix} 0 & 0 & 0 \\ 0 & 0 & -1 \\ 0 & 1 & 0 \end{pmatrix} \begin{pmatrix} 0 \\ 1 \\ 0 \end{pmatrix} = \begin{pmatrix} 0 \\ 0 \\ 1 \end{pmatrix}. \tag{3.21}$$

e(3)

Let E(3) denote the group of all Euclidean motions in three-dimensional space. So E(3) consists of all transformations of $\mathbb{R}^3$ of the form

$$\mathbf{r} \mapsto R\mathbf{r} + \mathbf{v}, \tag{3.22}$$

where $R \in O(3)$ and $\mathbf{v} \in \mathbb{R}^3$, in other words an orthogonal transformation followed by a translation. We can realize this transformation as matrix multiplication in four dimensions:

$$\begin{pmatrix} \mathbf{r} \\ 1 \end{pmatrix} \mapsto \begin{pmatrix} R & \mathbf{v} \\ 0 & 1 \end{pmatrix} \begin{pmatrix} \mathbf{r} \\ 1 \end{pmatrix}, \tag{3.23}$$

so E(3) can be thought of as the group of all 4×4 matrices of the form

$$\begin{pmatrix} R & \mathbf{v} \\ 0 & 1 \end{pmatrix}. \tag{3.24}$$

Therefore the Lie algebra e(3) can be identified with the algebra of all 4×4 matrices of the form

$$\begin{pmatrix} \mathbf{a} & \mathbf{b} \\ 0 & 0 \end{pmatrix}, \tag{3.25}$$

with $\mathbf{a} \in o(3)$ and $\mathbf{b} \in \mathbb{R}^3$. But recall from Eq. (3.20) that we have an identification of o(3) with $\mathbb{R}^3$. Therefore we can identify the six-dimensional algebra e(3) as a *vector space* with

$$o(3) \oplus o(3). \tag{3.26}$$

If $[\ ,\ ]_{o(3)}$ denotes the bracket in the Lie algebra o(3) then the Lie bracket in e(3), which is just the commutator of 4×4 matrices restricted to matrices in e(3) as above, is

$$[(\mathbf{a}_1, \mathbf{b}_1), (\mathbf{a}_2, \mathbf{b}_2)] = ([\mathbf{a}_1, \mathbf{a}_2]_{o(3)}, [\mathbf{a}_1, \mathbf{b}_2]_{o(3)} - [\mathbf{a}_2, \mathbf{b}_1]_{o(3)}). \qquad (3.27)$$

The bracket of two elements of the form $(\mathbf{a}, 0)$ is just the o(3) bracket in the $\mathbf{a}$ position, that is

$$[(\mathbf{a}, 0), (\mathbf{b}, 0)] = ([\mathbf{a}, \mathbf{b}]_{o(3)}, 0). \qquad (3.28)$$

The bracket of an $\mathbf{a}$ and a $\mathbf{b}$ is the o(3) bracket in the $\mathbf{b}$ position, that is

$$[(\mathbf{a}, 0), (0, \mathbf{b})] = (0, [\mathbf{a}, \mathbf{b}]_{o(3)}). \qquad (3.29)$$

Finally, the bracket of two $\mathbf{b}$s is zero:

$$[(0, \mathbf{b}_1), (0, \mathbf{b}_2)] = 0. \qquad (3.30)$$

o(4)

This is the algebra of all antisymmetric 4×4 matrices, so a basis is given by

$$\delta_1 = \begin{pmatrix} 0 & 0 & 0 & 0 \\ 0 & 0 & -1 & 0 \\ 0 & 1 & 0 & 0 \\ 0 & 0 & 0 & 0 \end{pmatrix}, \quad \delta_2 = \begin{pmatrix} 0 & 0 & 1 & 0 \\ 0 & 0 & 0 & 0 \\ -1 & 0 & 0 & 0 \\ 0 & 0 & 0 & 0 \end{pmatrix}, \quad \delta_3 = \begin{pmatrix} 0 & -1 & 0 & 0 \\ 1 & 0 & 0 & 0 \\ 0 & 0 & 0 & 0 \\ 0 & 0 & 0 & 0 \end{pmatrix},$$

$$\omega_1 = \begin{pmatrix} 0 & 0 & 0 & 1 \\ 0 & 0 & 0 & 0 \\ 0 & 0 & 0 & 0 \\ -1 & 0 & 0 & 0 \end{pmatrix}, \quad \omega_2 = \begin{pmatrix} 0 & 0 & 0 & 0 \\ 0 & 0 & 0 & 1 \\ 0 & 0 & 0 & 0 \\ 0 & -1 & 0 & 0 \end{pmatrix}, \quad \omega_3 = \begin{pmatrix} 0 & 0 & 0 & 0 \\ 0 & 0 & 0 & 0 \\ 0 & 0 & 0 & 1 \\ 0 & 0 & -1 & 0 \end{pmatrix},$$

$$(3.31)$$

and we write a general element of o(4) as $a_i \delta_i + b_i \omega_i, i = 1, 2, 3$, or as

$$\begin{pmatrix} \mathbf{a} & \mathbf{b} \\ -\mathbf{b}^T & 0 \end{pmatrix}, \qquad (3.32)$$

with $\mathbf{a} \in o(3)$ and $\mathbf{b} \in \mathbb{R}^3$. Once again, this six-dimensional Lie algebra can be identified as a vector space with $o(3) \oplus o(3)$. The bracket of two $\mathbf{a}$s is the same as before, as is the bracket of an $\mathbf{a}$ and a $\mathbf{b}$. What is new is the bracket of two $\mathbf{b}$s. In the algebra o(4) we have

$$[(0, \mathbf{b}_1), (0, \mathbf{b}_2)] = ([\mathbf{b}_1, \mathbf{b}_2]_{o(3)}, 0). \qquad (3.33)$$

$o(3,1)$

This is the algebra of 4×4 matrices of the form

$$\begin{pmatrix} \mathbf{a} & \mathbf{b} \\ \mathbf{b}^T & 0 \end{pmatrix}, \tag{3.34}$$

with $\mathbf{a} \in o(3)$ and $\mathbf{b} \in \mathbb{R}^3$. It can be shown that this is the Lie algebra of the Lorentz group — the group of all linear transformations of $\mathbb{R}^4$ that preserve the quadratic form

$$x_1^2 + x_2^2 + x_3^2 - x_4^2. \tag{3.35}$$

Once again the bracket between two **a**s and an **a** and a **b** is as before. But now

$$[(0, \mathbf{b}_1), (0, \mathbf{b}_2)] = (-[\mathbf{b}_1, \mathbf{b}_2]_{o(3)}, 0). \tag{3.36}$$

In all three cases, $e(3)$, $o(4)$, $o(3, 1)$, the Lie algebra decomposes as a vector space direct sum

$$\mathcal{G} = k + p, \tag{3.37}$$

with

$$[k, k] \subset k, \quad [k, p] \subset p, \quad \text{and} \quad [p, p] \subset k. \tag{3.38}$$

In these three cases $k = p = o(3)$ as vector spaces, and the $[k, k]$ and $[k, p]$ brackets are the $o(3)$ brackets. The bracket of p back into k is given by $\lambda [\ ,\ ]_{o(3)}$, where $\lambda = 1$ for $o(4)$, $\lambda = 0$ for $e(3)$, and $\lambda = -1$ for $o(3, 1)$.

It is notable that in the $o(4)$ case we have $k = p$ as vector spaces, and when we make this vector space identification all three brackets are the same: The subspaces

$$\Delta = \{(\mathbf{a}, \mathbf{a})\} \quad \text{the diagonal} \tag{3.39}$$

and

$$\Delta^T = \{(\mathbf{b}, -\mathbf{b})\} \quad \text{the antidiagonal} \tag{3.40}$$

are both subalgebras isomorphic to k and we have the *Lie algebra* direct sum decomposition

$$\mathcal{G} = \Delta \oplus \Delta^T. \tag{3.41}$$

Thus $o(4)$ is Lie algebra isomorphic to $o(3) \oplus o(3)$. This identification in Lie theory has remarkable physical consequences as will become apparent in Chapter 8.

3.3 Poisson Brackets

The implementation of symmetries is seen most clearly in the Hamiltonian formulation of classical mechanics. In this formulation an observable is a function on phase space. In addition to the usual addition and multiplication, the space of functions on phase space has another binary operation known as the *Poisson bracket*: Given two functions, f and g, their Poisson bracket $\{f, g\}$ is a third function on phase space, and this operation satisfies the following rules:

(a) $\{f, g\}$ is bilinear in f and g;
(b) $\{\ ,\ \}$ is antisymmetric, that is $\{g, f\} = -\{f, g\}$;
(c) Jacobi's identity: $\{f, \{g, h\}\} = \{\{f, g\}, h\} + \{g, \{f, h\}\}$; and
(d) f acting on gh satisfies the Leibniz rule: $\{f, gh\} = \{f, g\}h + g\{f, h\}$.

The first three axioms are just the axioms for a Lie algebra: The space of functions on phase space form a Lie algebra under the Poisson bracket. Axiom (d) ties the Poisson bracket in with usual multiplication. It says that the Poisson bracket acts as a *derivation*. A commutative algebra $\mathcal{A}$ having a bracket satisfying the above axioms is called a *Poisson algebra*.

For example, the phase space X for a single classical particle moving in ordinary three-dimensional space is the six-dimensional space consisting of all possible positions and momenta. So X is six-dimensional, with coordinates $(x_1, x_2, x_3, p_1, p_2, p_3)$. The Poisson brackets of these coordinate functions are

$$\{x_i, x_j\} = \{p_i, p_j\} = 0 \quad \text{and} \quad \{x_i, p_j\} = \delta_{ij}, \quad \text{with } i, j = 1, 2, 3. \quad (3.42)$$

When these bracket relations are realized we say that the coordinates p_1, p_2, p_3 are the *canonically conjugate* variables to x_1, x_2, x_3. Using the axioms we find

$$\{x_i, p_j^2\} = p_j\{x_i, p_j\} + \{x_i, p_j\}p_j = 2p_j\delta_{ij}, \quad (3.43)$$

and, by iteration,

$$\{x_i, p_j^n\} = np_j^{n-1}\delta_{ij}. \quad (3.44)$$

Therefore

$$\{x_i, g\} = \frac{\partial g}{\partial p_i}, \quad (3.45)$$

and similarly

$$\{p_i, g\} = -\frac{\partial g}{\partial x_i}, \quad (3.46)$$

for any polynomial $g = g(x, p)$. We then have

$$\{f, g\} = \frac{\partial f}{\partial x_i} \frac{\partial g}{\partial p_i} - \frac{\partial f}{\partial p_i} \frac{\partial g}{\partial x_i}, \tag{3.47}$$

for any pair of polynomials f and g. This is, in fact, the formula for the Poisson bracket of any pair of functions.

A transformation is called ϕ *symplectic* if the induced transformation on functions

$$f \mapsto f \circ \phi^{-1}, \tag{3.48}$$

preserves the Poisson bracket, that is, if

$$\{f, g\} \circ \phi^{-1} = \{f \circ \phi^{-1}, g \circ \phi^{-1}\} \tag{3.49}$$

for all functions f and g. It turns out that every function f on phase space generates a (local) one-parameter group, ϕ_t, of symplectic transformations such that

$$\{f, g\} = \left[\frac{d}{dt}\right]_{t=0} g \circ \phi_{-t}. \tag{3.50}$$

So forming the Poisson bracket with f (on the left) is the infinitesimal symmetry associated to f. A function on phase space serves a double role: It is an observable, and it also determines an infinitesimal symmetry of the space of observables via the Poisson bracket.

Let $\mathcal{G}$ be a Lie algebra. We say that we have a *Hamiltonian action* of $\mathcal{G}$ on X if we are given a Lie algebra homomorphism $\sigma : \mathcal{G} \mapsto \mathcal{A}$, where $\mathcal{A}$ denotes the space of functions on X. Thus σ assigns to each element $\xi \in \mathcal{G}$ a function $\sigma(\xi)$ on X and

$$\{\sigma(\xi), \sigma(\eta)\} = \sigma([\xi, \eta]) \tag{3.51}$$

for every $\eta \in \mathcal{G}$.

For example, consider the case of the phase space of a single free particle as above. Let $\mathcal{G}$ be the Lie algebra consisting of all 3×3 matrices under the commutator bracket, and define

$$\sigma(M) = -M_{ij} x_i p_j. \tag{3.52}$$

Equation (3.51) holds, since

$$\{\sigma(\xi), \sigma(\eta)\} = M_{ij} N_{mn} \left(\frac{\partial(x_i p_j)}{\partial x_k} \frac{\partial(x_m p_n)}{\partial p_k} - \frac{\partial(x_i p_j)}{\partial p_k} \frac{\partial(x_m p_n)}{\partial x_k}\right)$$

$$= M_{ij} N_{mn} (\delta_{ik} \delta_{nk} p_j x_m - \delta_{jk} \delta_{mk} x_i p_n)$$

$$= M_{ij}N_{mi}p_jx_m - M_{im}N_{mn}x_ip_n$$

$$= (NM)_{ij}x_ip_j - (MN)_{ij}x_ip_j = -[M,N]_{ij}x_ip_j = \sigma([\xi,\eta]).$$

$$(3.53)$$

Notice also that

$$\{\sigma(M), x_k\} = M_{ik}x_i \quad \text{and} \quad \{\sigma(M), p_k\} = -M_{kj}p_j. \tag{3.54}$$

So if we identify the function x_k with the kth basis element in $\mathbb{R}^3$ then the first equation in (3.54) identifies the bracket with $\sigma(M)$ with the matrix M_{ij} and so σ defines an isomorphism between the Lie algebra gl(3, $\mathbb{R}$) and the subalgebra of $\mathcal{A}$ consisting of all homogeneous polynomials of degree two, which are of degree one in x and degree one in p. For example, if $L_3 = x_1p_2 - x_2p_1$ then

$$\{L_3, x_1\} = x_2, \quad \{L_3, x_2\} = -x_1, \quad \{L_3, x_3\} = 0, \tag{3.55}$$

and

$$\{L_3, p_1\} = p_2, \quad \{L_3, p_2\} = -p_1, \quad \{L_3, p_3\} = 0. \tag{3.56}$$

The first of these equations says that L_3 acts on $\mathbf{r}$ as an infinitesimal rotation about the x_3-axis. Let o(3)$\in$ gl(3, $\mathbb{R}$) be the Lie algebra of infinitesimal rotations and let L denote the restriction of σ to o(3). Recall that δ_3 denotes infinitesimal rotation about the x_3-axis (with similar notation for δ_1 and δ_2) and then $L_3 = L(\delta_3)$.

Let us use $\times$ to denote the vector product on $\mathbb{R}^3$. We have made the identification of o(3) with $\mathbb{R}^3$ so that the Lie bracket is given by $\times$. That is, the elements of o(3) are regarded as vectors and the Lie bracket becomes the vector product. The elements of δ_i then become the standard basis vectors. If we form the vector-valued function

$$\boldsymbol{L} = \mathbf{r} \times \boldsymbol{p} \tag{3.57}$$

then under the identification of o(3) with $\mathbb{R}^3$ we have, for any $\xi \in \mathbb{R}^3$,

$$L(\xi) = \boldsymbol{L} \cdot \xi, \tag{3.58}$$

and (3.51) becomes

$$L([\xi, \eta]) = \{L(\xi), L(\eta)\}. \tag{3.59}$$

The function $\boldsymbol{L}$ is called the *angular momentum*.

The time evolution of a mechanical system is the one-parameter group generated by the Hamiltonian H. The infinitesimal change in any observable f

under this time evolution is thus given by $\{H, f\}$. We say that H is invariant under the action σ of the Lie algebra $\mathcal{G}$ if

$$\{\sigma(\xi), H\} = 0 \tag{3.60}$$

for all $\xi \in \mathcal{G}$. This, by the antisymmetry of $\{\ ,\ \}$, implies that $\{H, \sigma(\xi)\} = 0$. In other words, all the functions $\sigma(\xi)$ are conserved quantities under the flow generated by H. Thus if H is rotationally invariant, this means that all the functions $L(\xi)$, or, what amounts to the same thing, each of the functions L_1, L_2, L_3 are constant along the trajectories of H. In other words, the vector-valued function L is constant. This is *the principle of conservation of angular momentum.*

Infinitesimal translation in the x_1 direction acts on phase space by the Poisson bracket with p_1. Indeed,

$$\{p_1, x_1\} = -1, \quad \{p_1, x_2\} = \{p_1, x_3\} = \{p_1, p_i\} = 0, \quad i = 1, 2, 3. \tag{3.61}$$

Thus invariance of the Hamiltonian under the full translation group implies conservation of the vector-valued function p. This is *the principle of the conservation of linear momentum.*

Thus invariance of the Hamiltonian under the full group of Euclidean motions containing all translations and rotations implies conservation of both linear and angular momentum. One of Galileo's principal revolutionary achievements was to replace the geocentric Aristotelian theory, which has an O(3) symmetry centered at the origin of the Earth, with the larger symmetry group E(3). Indeed, it is easy to check that the map

$$(\mathbf{a}, \mathbf{b}) \mapsto \mathbf{a} \cdot L + \mathbf{b} \cdot p \tag{3.62}$$

gives a Hamiltonian action of e(3) on phase space.

3.4 The Inverse Square Law

In the problem of a particle moving under the influence of a central force we no longer have translational symmetry, but we retain rotational symmetry about the origin. We do not have $\{H, p\} = 0$, but if the force is inversely proportional to the square of the distance we have the three conserved quantities $\mathcal{A}$ of Eq. (2.22) in addition to the quantities L. In the present formalism a quantity is conserved if its Poisson bracket with the Hamiltonian vanishes. We check that this is indeed the case. With

$$H = \frac{p^2}{2m} - \frac{k}{r}, \tag{3.63}$$

we have

$$\{H, \mathcal{A}_i\} = -\frac{1}{m}\epsilon_{ijk}\{1/r, p_j\}L_k - \frac{1}{m}p_j\{p_j, x_i/r\}$$

$$= \frac{1}{m}(\epsilon_{ijk}x_j L_k/r^3 + p_i/r - x_i p_j x_j/r^3), \qquad (3.64)$$

Using Eq. (3.46), since $\{L_i, f(r)\} = 0$ for f an arbitrary function of r. The first term is

$$\frac{1}{m}\epsilon_{ijk}\epsilon_{ksn}x_j x_s p_n/r^3 = \frac{1}{m}(\delta_{is}\delta_{jn} - \delta_{in}\delta_{js})x_j x_s p_n/r^3, \qquad (3.65)$$

which cancels the second and third terms. We are using

$$L_i = \epsilon_{ijk}x_j p_k. \qquad (3.66)$$

Exercise 3.1 Prove $\epsilon_{ijk}\epsilon_{imn} = \delta_{jm}\delta_{kn} - \delta_{jn}\delta_{km}$.

Exercise 3.2 Prove $\epsilon_{ijk}\epsilon_{ijm} = 2\delta_{km}$.

We want to see if these six quantities close to a Lie algebra. We still have

$$\{L(\xi), L(\eta)\} = L([\xi, \eta]) \qquad (3.67)$$

as before. We also have

$$\{L(\xi), \mathcal{A}(\eta)\} = \mathcal{A}([\xi, \eta]), \qquad (3.68)$$

which just says that $\mathcal{A}$ transforms as a vector. Indeed,

$$\{L_i, \mathcal{A}_j\} = \frac{1}{mk}\epsilon_{jsn}\{L_i, p_s L_n\} - \{L_i, x_j/r\}$$

$$= \frac{1}{mk}\epsilon_{jsn}(\{L_i, p_s\}L_n + p_s\{L_i, L_n\}) - \{L_i, x_j\}\frac{1}{r}$$

$$= \epsilon_{ijk}\left(\frac{(\mathbf{p} \times \mathbf{L})_k}{mk} - \frac{x_k}{r}\right) = \epsilon_{ijk}\mathcal{A}_k, \qquad (3.69)$$

We now wish to compute $\{\mathcal{A}_i, \mathcal{A}_j\}$. It is a somewhat lengthy calculation, but we go through it in detail because of the importance of the result, which will be needed in Chapter 8. We first calculate some Poisson brackets, which are used below:

$$\begin{aligned}
(mk)\{p_i, \boldsymbol{A}_j\} &= \epsilon_{jsn}\{p_i, p_s L_n\} - mk\{p_i, x_j/r\} \\
&= \epsilon_{jsn} p_s\{p_i, L_n\} - mk\{p_i, x_j\}/r - mkx_j\{p_i, 1/r\} \\
&= -\epsilon_{jsn}\epsilon_{niq}p_s p_q + mk\delta_{ij}/r - mkx_j x_i/r^3 \\
&= (\delta_{jq}\delta_{si} - \delta_{ji}\delta_{sq})p_s p_q + mk\delta_{ij}/r - mkx_i x_j/r^3 \\
&= p_i p_j - \mathbf{p}^2\delta_{ij} + mk\delta_{ij}/r - mkx_i x_j/r^3,
\end{aligned} \qquad (3.70)$$

and

$$\begin{aligned}
(mk)\{x_i/r, \boldsymbol{A}_j\} &= \epsilon_{jsn}\{x_i/r, p_s L_n\} = \epsilon_{jsn}(\{x_i, p_s L_n\}/r + \{1/r, p_s L_n\}x_i) \\
&= \epsilon_{jsn}(p_s\{x_i, L_n\}/r + \{x_i, p_s\}L_n/r + \{1/r, p_s\}L_n x_i) \\
&= \epsilon_{jsn}(-\epsilon_{nik}p_s x_k/r + \delta_{is}L_n/r - x_s x_i L_n/r^3) \\
&= -(\delta_{ji}\delta_{sk} - \delta_{jk}\delta_{si})p_s x_k/r + \epsilon_{jin}L_n/r - \epsilon_{jsn}x_s x_i L_n/r^3 \\
&= p_i x_j/r - \delta_{ij}x_m p_m/r - \epsilon_{ijm}L_m/r - \epsilon_{jmn}x_i x_m L_n/r^3.
\end{aligned}$$
$$(3.71)$$

The latter yields, remembering that $\mathbf{r} \cdot \mathbf{L} = 0$, and using Exercises 3.1 and 3.2,

$$(mk)^2\epsilon_{ijk}\{x_i/r, \boldsymbol{A}_j\} = -2mkL_k/r. \qquad (3.72)$$

We are now ready to do the main calculation:

$$\begin{aligned}
(mk)^2\epsilon_{ijk}\{\boldsymbol{A}_i, \boldsymbol{A}_j\} &= (mk)\epsilon_{ijk}(\epsilon_{isn}\{p_s L_n, \boldsymbol{A}_j\} - (mk)\{x_i/r, \boldsymbol{A}_j\}) \\
&= (mk)\epsilon_{ijk}(\epsilon_{isn}(p_s\{L_n, \boldsymbol{A}_j\} + \{p_s, \boldsymbol{A}_j\}L_n) \\
&\quad - (mk)\{x_i/r, \boldsymbol{A}_j\}).
\end{aligned} \qquad (3.73)$$

We evaluate the separate terms:

$$\begin{aligned}
\epsilon_{ijk}\epsilon_{isn}p_s\{L_n, \boldsymbol{A}_j\} &= \epsilon_{ijk}\epsilon_{isn}\epsilon_{njq}p_s\boldsymbol{A}_q \\
&= \epsilon_{ijk}(\delta_{ij}\delta_{sq} - \delta_{iq}\delta_{sj})p_s\boldsymbol{A}_q = \epsilon_{ijk}p_i\boldsymbol{A}_j,
\end{aligned} \qquad (3.74)$$

and

$$\begin{aligned}
(mk)\epsilon_{ijk}p_i\boldsymbol{A}_j &= \epsilon_{ijk}\epsilon_{jsq}p_i p_s L_q - (mk)\epsilon_{ijk}p_i x_j/r \\
&= (\delta_{iq}\delta_{ks} - \delta_{is}\delta_{kq})p_i p_s L_q + (mk)L_k/r \\
&= (-\mathbf{p}^2 + mk/r)L_k.
\end{aligned} \qquad (3.75)$$

Also

$$(mk)\epsilon_{ijk}\epsilon_{isn}\{p_s, \boldsymbol{A}_j\}L_n = \epsilon_{ijk}\epsilon_{isn}(p_s p_j - \mathbf{p}^2\delta_{sj} + mk\delta_{sj}/r$$
$$- mkx_s x_j/r^3)L_n$$
$$= (\delta_{js}\delta_{kn} - \delta_{jn}\delta_{ks})(p_s p_j - \mathbf{p}^2\delta_{sj} + mk/r\delta_{sj}$$
$$- mkx_s x_j/r^3)L_n$$
$$= (-2p^2 + 2mk/r)L_k - (-\mathbf{p}^2 + mk/r)L_k$$
$$= (-\mathbf{p}^2 + mk/r)L_k. \tag{3.76}$$

Combining these results, we get

$$(mk)^2\epsilon_{ijk}\{\boldsymbol{A}_i, \boldsymbol{A}_j\} = [(-\mathbf{p}^2 + mk/r) + (-\mathbf{p}^2 + mk/r) + 2mk/r]L_k$$
$$= -2(\mathbf{p}^2 - 2mk/r)L_k = -4mHL_k. \tag{3.77}$$

Multiply by ϵ_{ijk} to finally get

$$(mk)^2\epsilon_{ijk}\epsilon_{mnk}\{\boldsymbol{A}_m, \boldsymbol{A}_n\} = 2(mk)^2\{\boldsymbol{A}_i, \boldsymbol{A}_j\} = -4m\epsilon_{ijk}HL_k, \tag{3.78}$$

and

$$\{\boldsymbol{A}_i, \boldsymbol{A}_j\} = \epsilon_{ijk}\left(\frac{-2H}{mk^2}\right)L_k. \tag{3.79}$$

The occurrence of $H(\mathbf{r}, \mathbf{p})$ in the last equation prevents the quantities L_i and $\boldsymbol{A}_i$ from closing to a Lie algebra on the whole phase space. But when the motion is in a fixed orbit $H(\mathbf{r}, \mathbf{p}) = E$, with E a constant energy. We can then divide the phase space into three sets, the two open regions where $H > 0$ and $H < 0$, and their common boundary where $H = 0$. In the region where $H > 0$ we can rescale $\boldsymbol{A}$ by $2H/mk^2$ to get an o(3, 1) algebra. In the region where $H < 0$ we can rescale by $-2H/mk^2$ to get an o(4) algebra.

That is, we define

$$A = \mathcal{N}\boldsymbol{A}, \tag{3.80}$$

where

$$\mathcal{N}^{-2} = -\frac{2H}{mk^2}, \tag{3.81}$$

to get

$$\{L_i, L_j\} = \epsilon_{ijk}L_k, \quad \{L_i, A_j\} = \epsilon_{ijk}A_k, \quad \{A_i, A_j\} = \epsilon_{ijk}L_k. \tag{3.82}$$

This shows that the quantities L_i and A_i close to form an o(4) Lie algebra.

We can use the Lie algebra isomorphism of o(4) with o(3)⊕ o(3) to form the direct sum of two o(3) algebras. Define

$$M_i = L_i + A_i , \quad N_i = L_i - A_i. \tag{3.83}$$

Then the Poisson structure becomes

$$\{M_i, M_j\} = \epsilon_{ijk} M_k, \quad \{N_i, N_j\} = \epsilon_{ijk} N_k, \quad \{M_i, N_j\} = 0. \tag{3.84}$$

We have shown that a force which is inversely proportional to the square of the distance between the particles leads, in the case of bound states, to six conserved quantities which form, when properly normalized, the generators of an so(4) algebra that acts on the functions of phase space. This result, which appears here in a somewhat abstract form, gives rise to physical results when applied to the spectrum of hydrogen in Chapter 8.

Notes on Chapter 3

This review of Hamiltonian mechanics is standard, and can be found in any good text on classical mechanics. It is included here to make the treatment self-contained. The same is true for the relation between Lie groups and Lie algebras. More information can be found in Weaver and Sattinger [WS10], or Marsden and Ratiu [MR99]. We follow here the treatment of Guillemin and Sternberg [GS90]. The calculation of the Poisson bracket between the components of the Laplace vector is the classical version of the calculation of the corresponding commutator by Sudbery [Sud86].

Chapter 4
From Solar Systems to Atoms

4.1 Rutherford Scattering

"It was quite the most incredible event that has ever happened to me in my life. It was almost as incredible as if you fired a 15-inch shell at a piece of tissue paper and it came back and hit you."

Lord Rutherford, quoted in *Rutherford and the Nature of the Atom*, E. N. da C. Andrade, Heinemann, London, 1964.

In 1897 J. J. Thompson measured the mass-to-charge ratio of the electron. In 1902 Lord Kelvin proposed that the atom consisted of a sphere of positive charge in which the electrons were embedded. In 1906 W. Thompson reasoned that the number of electrons in an atom was approximately equal to the atom's atomic weight. On this basis, there would be only one electron in the hydrogen atom.

In 1911 Rutherford [Rut11] established, from the data of scattering experiments with alpha particles impinging on gold folie, that the positive charge in the atom is concentrated in a nucleus 10,000 times smaller than the atom itself. This nucleus also carries most of the mass of the atom. This was the background on which Bohr [Boh13] developed his model of the hydrogen atom in 1913. The electron and its nucleus form a two-body system, with an electron mass that is negligible in comparison to the mass of the nucleus. Hence one has a point electron moving in the Coulomb potential of a nucleus of charge one.

To understand Rutherford's reasoning we shall derive his formula for scattering. In this derivation we shall again make use of the Laplace vector of the system. We do this in order to get some practice with the application of the Laplace vector, and also because the formula is of paramount importance on the road to understanding atomic physics.

4.2 Conservation of the Laplace Vector

We consider impinging particles of charge e, and as target a nucleus of charge Ze. The central force is repulsive, and k in Eq. (2.17) is negative, $k = -|k| = -Z\alpha$. Then the Laplace vector is

$$\mathcal{A} = -\frac{(\mathbf{p} \times \mathbf{L})}{Z\alpha m} - \hat{\boldsymbol{r}}. \tag{4.1}$$

The scattering process is described as in Figure 4.1. Long before the collision takes place the initial impulse of a particle in the beam is $\mathbf{p}_i = mv_0\mathbf{e}_i$, where v_0 is the velocity of the incoming particle and $\mathbf{e}_i$ is a unit vector in the direction of the incoming velocity. The final velocity is $\mathbf{p}_f = mv_0\mathbf{e}_f$, with the same velocity v_0 (the process is *elastic*) and $\mathbf{e}_f$ is a unit vector in the direction of the outgoing velocity. $\mathbf{L}_i = \mathbf{L}_f = mv_0b\hat{\boldsymbol{k}}$, where $\mathbf{L}$ is the conserved angular momentum, and $\hat{\boldsymbol{k}}$ is a unit vector pointing up through the page, perpendicular to the scattering plane. b is the *impact parameter*, the distance that the incoming particle would pass by the point target in the absence of interaction. The initial incoming direction is $\hat{\boldsymbol{r}}_i = -\mathbf{e}_i$, and the final outgoing direction $\hat{\boldsymbol{r}}_f = \mathbf{e}_f$.

The initial Laplace vector is

$$\mathcal{A}_i = -\left(\frac{mv_0^2}{Z\alpha}\right)b\mathbf{n}_i + \mathbf{e}_i, \tag{4.2}$$

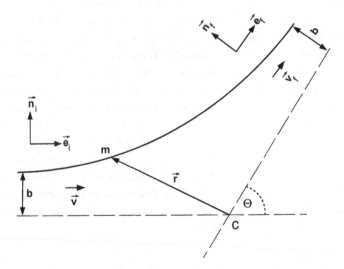

Figure 4.1.   The incoming and outgoing directions and the scattering angle.

where $\mathbf{n}_i$ is the unit vector in Figure 4.1 perpendicular to $\mathbf{e}_i$ and in the scattering plane. The final Laplace vector is

$$\mathcal{A}_f = - \left(\frac{mv_0^2}{Z\alpha} \right) b\mathbf{n}_f - \mathbf{e}_f, \qquad (4.3)$$

where $\mathbf{n}_f$ is the unit vector perpendicular to $\mathbf{e}_f$ and in the scattering plane.

Conservation of the Laplace vector yields

$$\mathcal{A}_i \cdot \mathbf{e}_i = \mathcal{A}_f \cdot \mathbf{e}_i. \qquad (4.4)$$

With $(\mathbf{e}_f \cdot \mathbf{e}_i) = \cos\theta$, where θ is the scattering angle, and $(\mathbf{n}_f \cdot \mathbf{e}_i) = \cos(\pi/2 + \theta)$ $= -\sin\theta$, this leads to

$$Z\alpha = mv_0^2 b \sin\theta - Z\alpha \cos\theta. \qquad (4.5)$$

Rearranging this gives a relation between b and θ,

$$\tan(\theta/2) = \frac{Z\alpha}{mv_0^2 b}, \qquad (4.6)$$

and this enables us to calculate the scattering angle for any impact parameter.

4.3 The Differential Cross Section

The differential scattering cross section is defined in terms of the number of particles scattered into the shaded band $2\pi d\theta$ in Figure 4.2. From the figure this is the same as the number of particles incident on the area $2\pi b\, db$, and the differential cross

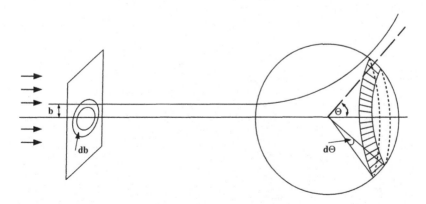

Figure 4.2. Scattering of an incident beam of particles by a center of force.

section is

$$\frac{d\sigma}{d\Omega} = \frac{b}{\sin\theta}\left|\frac{db}{d\theta}\right|. \tag{4.7}$$

We calculate, from Eq. (4.6),

$$\left|\frac{db}{d\theta}\right| = \frac{Z\alpha}{2mv_0^2\sin^2(\theta/2)}. \tag{4.8}$$

Therefore we obtain *Rutherford's scattering formula*:

$$\frac{d\sigma}{d\Omega} = \frac{(Z\alpha)^2}{4(mv_0^2)^2\sin^4(\theta/2)}. \tag{4.9}$$

The result for the Thompson atom, which involves multiple scattering, decreases much more rapidly then the characteristic $\sin^{-4}(\theta/2)$ of Rutherford's formula. This is the reason for Rutherford's surprise, noted earlier. Together with his associates, Geiger and Marsden, he verified the formula in great detail, including its proportionality to Z^2, and the inverse squared proportionality to the kinetic energy of the incoming particles. The formula was also found to be accurate down to small distances, and the finite size of the nucleus, at which deviations from the formula were observed, was around $10^{-14}m = 10$ Fermi for an aluminium foil. The success of the formula proved the correctness of the assumptions on which it is based, and established the solar system model for atoms.

Notes on Chapter 4

Rutherford scattering is analyzed using the Laplace vector by Basano and Bianchi [BB80], and in textbooks on classical mechanics; see, for example, Poole, Goldstein and Safko [PGS01].

Chapter 5
The Bohr Model

"The spectrum of the hydrogen atom has proved to be the Rosetta stone of modern physics."

Theodor W. Hänsch, Arthur L. Schawlow, and George W. Series, in *Rev. Mod. Phys.* 54, p. 697 (1982).

5.1 Spectroscopic Series

Balmer published in 1885 an empirical formula for the frequencies of the spectral lines of the hydrogen spectrum in the visible region,

$$h\nu = R_\infty \left(\frac{1}{4} - \frac{1}{N^2} \right), \tag{5.1}$$

with $N = 3, 4, 5, 6$. We have used here a modern notation for the constants, with h the constant later discovered by Planck [Pla00], ν the frequency of the light emitted, and R_∞ the Rydberg constant.

This provided the key for the Bohr model of the hydrogen atom. Subsequently the Balmer formula was generalized to further series, these were found *after* Bohr had proposed his model; see Table 5.1 and Figure 5.1.

5.2 The Postulates of the Model

In 1913 Niels Bohr presented a model of the hydrogen atom [Boh13], which quantitatively explained the spectroscopic measurements to the accuracy which was obtainable at that time. He based his model on the following postulates:

1. An electron in a hydrogen atom moves under the influence of an attractive Coulomb potential between the electron and the nucleus according to the laws of classical mechanics (i.e., in particular in circular orbits).
2. Quantization condition: Instead of the infinity of orbits possible in classical mechanics, it is only possible for an electron to move in an orbit for which its orbital angular momentum L is an integral number.

Table 5.1.   The hydrogen series

Names	Wavelength ranges	Formulas	
Lyman	Ultraviolet	$\nu = R_\infty \left(\frac{1}{1^2} - \frac{1}{N^2} \right)$	$N = 2, 3, 4, \ldots$
Balmer	Near Ultraviolet and Visible	$\nu = R_\infty \left(\frac{1}{2^2} - \frac{1}{N^2} \right)$	$N = 3, 4, 5, \ldots$
Paschen	Infrared	$\nu = R_\infty \left(\frac{1}{3^2} - \frac{1}{N^2} \right)$	$N = 4, 5, 6, \ldots$
Brakett	Infrared	$\nu = R_\infty \left(\frac{1}{4^2} - \frac{1}{N^2} \right)$	$N = 5, 6, 7, \ldots$
Pfund	Infrared	$\nu = R_\infty \left(\frac{1}{5^2} - \frac{1}{N^2} \right)$	$N = 6, 7, 8, \ldots$

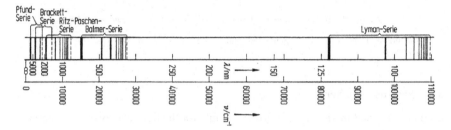

Figure 5.1.   The spectral series of hydrogen.

3. Despite the fact that it is constantly accelerating, an electron moving in such an allowed orbit does not radiate electromagnetic energy through bremsstrahlung. Thus, its total energy E remains constant.

4. Electromagnetic radiation is emitted if an electron, initially moving in an orbit of total energy E_i, discontinuously changes its motion so that it moves in an orbit of total energy E_f. The frequency of the emitted radiation ν is equal to the quantity $(E_i - E_f)$ divided by Planck's constant $2\pi (= 2\pi\hbar)$.

The first postulate bases the model on the picture introduced by Rutherford. The second postulate introduces quantization of orbital angular momentum of an atomic electron moving under the influence of an inverse square (Coulomb) force:

$$L = N, \quad N = 1, 2, 3, \ldots \tag{5.2}$$

Planck's quantization condition for the *energy* of a particle, such as an electron, executing simple harmonic motion under the influence of a harmonic restoring force, was $E = 2\pi N\nu$, $N = 0, 1, 2, 3, \ldots$, which is at first sight different from Bohr's condition. The third postulate removes the problem of the instability of an electron moving in a circular orbit, due to the emission of the electromagnetic radiation required of the electron by classical electromagnetic theory, by

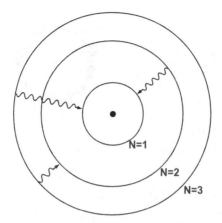

Figure 5.2. The Bohr model of the hydrogen atom.

postulating that this particular feature of the classical theory is not valid for the case of an atomic electron. The postulate was based on the fact that atoms are observed by experiment to be stable — even though this is not predicted by classical theory. The fourth postulate, $2\pi\nu = E_i - E_f$, is really just Einstein's postulate that the frequency of a photon of electromagnetic radiation is equal to the energy carried by the photon divided by Planck's constant; see Figure 5.2.

5.3 The Predictions of the Model

Consider an atom consisting of a nucleus of charge $+Ze$ and mass taken to be infinite, compared to the mass of the electron. For a neutral hydrogen atom $Z = 1$, for a singly ionized helium atom $Z = 2$, for a doubly ionized lithium atom $Z = 3$, etc. We assume that the electron revolves in a circular orbit about the nucleus, which, being infinitely heavy, remains fixed in space. The condition of mechanical stability of the electron is

$$\frac{Z\alpha}{r^2} = \frac{mv^2}{r}, \tag{5.3}$$

where $\alpha = e^2/4\pi$ is the fine structure constant, v is the speed of the electron in its orbit, and r is the radius of the orbit. The left-hand side of this equation is the Coulomb force acting on the electron, and the right-hand side is ma, where a is the centripetal acceleration of the electron in its circular orbit.

Now, the orbital angular momentum of the electron, $L = mvr$, is a constant, because the force acting on the electron is entirely in the radial direction. Applying

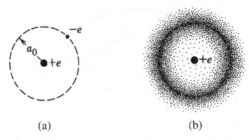

(a)   (b)

Figure 5.3.   The electron in the ground state of the Bohr atom. (a) Classical picture; (b) Quantum Mechanical picture.

the quantization condition, Eq. (5.2), to L, we get

$$mvr = N, \quad N = 1, 2, 3, \ldots \tag{5.4}$$

Solving for v and substituting into Eq. (5.3), we obtain

$$Z\alpha = mv^2 r = mr \left(\frac{N}{mr}\right)^2 = \frac{N^2}{mr}, \tag{5.5}$$

so

$$r = \frac{N^2}{Z\alpha m}, \quad N = 1, 2, 3, \ldots \tag{5.6}$$

and

$$v = \frac{N}{mr} = \frac{Z\alpha}{N}, \quad N = 1, 2, 3, \ldots \tag{5.7}$$

The application of the angular momentum quantization condition has restricted the possible circular orbits to those of radii given by Eq. (5.6). Note that these radii are proportional to the square of the *quantum number N*. If we evaluate the radius of the smallest orbit ($N = 1$) for a hydrogen atom ($Z = 1$) by inserting the known values of m and α, we obtain $r = a_0 = 1/\alpha m_e \approx 0.5 \, \mathring{A}$, the *Bohr radius*. We shall show later that the electron has its minimal total energy when in the orbit corresponding to $N = 1$. Consequently we may interpret the radius of this orbit as a measure of the radius of the hydrogen atom in its ground state. It is in good agreement with the estimate that the order of magnitude of an atomic radius is $1 \, \mathring{A}$. Hence, Bohr's postulate predicts a reasonable size for the atom; see Figures 5.3 and 5.4.

Evaluating the orbital velocity of an electron in the ground state of hydrogen, we find $v = 2.2 \times 10^6 \, m/sec$. It is apparent from the equation that this is the largest velocity possible for a hydrogen atom electron. The fact that this velocity is less than 1% of the velocity of light is the justification for using non-relativistic mechanics

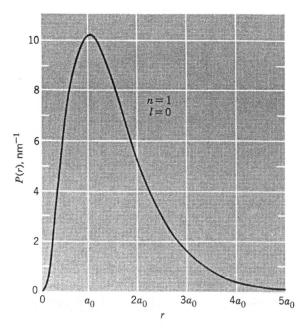

Figure 5.4. The probability of finding the electron at a given distance from the nucleus.

in the Bohr model. On the other hand, Eq. (5.7) shows that for large values of Z the electron velocity becomes relativistic, and the model cannot be applied.

Next we calculate the total energy of an atomic electron moving in one of the allowed orbits. Let us define the potential energy to be zero when the electron is at rest infinitely distant from the nucleus. Then the potential energy V at any finite distance r can be obtained by integrating the work that would be done by the Coulomb force acting from r to ∞. Thus

$$V = \int_r^\infty \frac{Z\alpha}{r^2} dr = -\frac{Z\alpha}{r}. \tag{5.8}$$

The potential energy is negative because the Coulomb force is attractive; it takes work to move the electron from r to ∞ against this force. The kinetic energy of the electron, E_{kin}, can be evaluated, with the aid of Eq. (5.3), to be

$$E_{kin} = \frac{1}{2}mv^2 = \frac{Z\alpha}{2r}. \tag{5.9}$$

The total energy of the electron, E, is then

$$E = E_{kin} + V = -\frac{Z\alpha}{2r} = -E_{kin}, \tag{5.10}$$

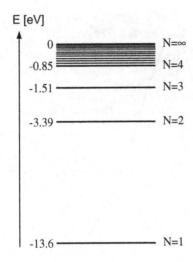

Figure 5.5.   Energy-level diagram of the Bohr model.

in agreement with the virial theorem. Using Eq. (5.6) for r in the preceding equation, we have

$$E = -\left(\frac{m}{2}\right)\frac{(Z\alpha)^2}{N^2}, \quad N = 1, 2, 3, \dots \tag{5.11}$$

We see that the quantization of the orbital angular momentum of the electron leads to a quantization of its total energy.

The information contained in Eq. (5.11) is presented as an energy-level diagram in Figure 5.5. The energy of each level, as evaluated from Eq. (5.11) is shown on the left, in terms of electron volts, and the quantum number of the level is shown on the right. The diagram is so constructed that the distance from any level to the level of zero energy is proportional to the energy of that level. Note that the lowest allowed total energy occurs for the smallest quantum number $N = 1$. As N increases, the total energy of the quantum states becomes less negative, with E approaching zero as N approaches infinity. Since the state of lowest total energy is the ground state, the normal state of the electron in a one-electron atom is the state for which $N = 1$.

The energy binding the electron to the nucleus may be calculated from Eq. (5.11). The binding energy is numerically equal to the energy of the ground state, corresponding to $N = 1$. This yields, with $Z = 1$, $E = -(m/2)\alpha^2 = -13.6\,eV$, which agrees very well with the experimentally observed binding energy for hydrogen. Next we calculate the frequency ν of the electromagnetic radiation emitted when the electron makes a transition from the quantum state N_i

to the quantum state N_f, that is, when an electron initially moving in an orbit characterized by the quantum number N_i discontinuously changes its motion so that it moves in an orbit characterized by quantum number N_f. Using Bohr's fourth postulate, Eq. (5.2), and Eq. (5.11), we have

$$\nu = \frac{E_i - E_f}{2\pi} = \frac{Z^2 \alpha^2 m}{4\pi} \left(\frac{1}{N_f^2} - \frac{1}{N_i^2} \right) = R_\infty Z^2 \left(\frac{1}{N_f^2} - \frac{1}{N_i^2} \right), \quad (5.12)$$

with

$$R_\infty = \frac{\alpha^2 m}{4\pi}, \quad (5.13)$$

and where N_i and N_f are integers.

The essential predictions of the Bohr model are contained in Eqs. (5.11) and (5.12). We discuss the emission of electromagnetic radiation by a one-electron Bohr atom in terms of these equations:

1. The atom is normally in its ground state $N = 1$.
2. In an electric discharge, or in some other process, the atom receives energy due to collisions, etc. This means that the electron must make a transition to an *excited state*, in which $N > 1$.
3. Obeying the common tendency of all physical systems, the atom will emit its excess energy and return to the ground state. This is accomplished by a series of transitions in which the electron drops to excited states of successively lower energy, finally reaching the ground state. In each transition electromagnetic radiation is emitted with a wavelength that depends on the energy lost by the electron, i.e., on the initial and final quantum numbers.
4. In the very large number of excitation and de-excitation processes that take place during a measurement of an atomic spectrum, all possible transitions occur and the complete spectrum is emitted. The reciprocal wavelengths of the set of lines that constitute the spectrum are given by Eq. (5.12), where we allow N_i and N_f to take on all possible integral values subject only to the restriction that $N_i > N_f$.

According to the Bohr model, each of the five known series of the hydrogen spectrum arises from a subset of transitions in which the electron goes to a certain final quantum state N_f. For the Lyman series $N_f = 1$, for the Barmer series $N_f = 2$, for the Paschen series $N_f = 3$, for the Brakett series $N_f = 4$, and for the Pfund series $N_f = 5$. The wavelength of the lines of all these series are fitted very accurately by Eq. (5.12) by using the appropriate value of N_f. This was a great triumph of the Bohr model. The success of the model was particularly impressive because the Lyman, Brakett and Pfund series had not been discovered at the time the model was developed by Bohr. The existence of these series was predicted, and the series were soon found experimentally by the persons after whom they were named.

The model worked equally well when applied to the case of one-electron atoms with $Z = 2$, i.e., singly ionized helium atoms He^+. Such atoms can be produced by passing a spark through normal helium gas. They make their presence apparent by emitting a simpler spectrum than that emitted by normal helium atoms. In fact, the spectrum of He^+ is exactly the same as the hydrogen spectrum except that the reciprocal wavelengths of all the lines are almost exactly four times as great. This is explained very easily, in terms of the Bohr model, by setting $Z^2 = 4$ in Eq. (5.11).

The properties of the absorption spectrum of one-electron atoms are also easy to understand in terms of the Bohr model. Since the atomic electron must have a total energy exactly equal to the energy of one of the allowed energy states, the atom can only absorb discrete amounts of energy from the incident electromagnetic radiation. This fact leads to the idea that we consider the incident radiation to be a beam of photons, and that only those photons can be absorbed whose frequency is given by $E = 2\pi\nu$, where E is one of the discrete amounts of energy which can be absorbed by the atom. The process of absorbing electromagnetic radiation is then just the inverse of the normal emission process, and the lines of the absorption spectrum will have exactly the same wavelengths as the lines of the emission spectrum. Normally the atom is always initially in the ground state $N = 1$, so only absorbtion processes from $N = 1$ to $N > 1$ can occur. Thus, the only absorption lines which correspond (for hydrogen) to the Lyman series will normally be observed. However, if the gas containing the absorbing atoms is at a very high temperature, then, owing to collisions, some of the atoms will initially be in the first excited state $N = 2$, and absorption lines corresponding to the Balmer series will be observed. Balmer absorbtion lines are actually observed in the hydrogen gas of some stellar atmospheres. This gives us a way of estimating the temperature of the surface of a star.

5.4 Correction for Finite Nuclear Mass

In the previous section (5.3) we assumed the mass of the atomic nucleus to be infinitely large compared to the mass of the atomic electron, so that the nucleus remains fixed in space. This is a good approximation even for hydrogen, which contains the lightest nucleus, since the mass of the nucleus is about 2000 times larger than the electron mass. However, the spectroscopic data are so very accurate that before we make a comparison of these data with the Bohr model we must take into account the fact that the nuclear mass is actually finite. In such a case the electron and the nucleus move about their common center of mass. However, it is not difficult to show that in such a system the electron moves relative to the nucleus as though the nucleus were fixed and the mass of the electron were slightly reduced to the value μ, the *reduced mass* of the system. The equations of motion of the system are the same as those we have considered if we simply substitute μ for m, where

$$\mu = \frac{mM}{m + M},\tag{5.14}$$

is less than m by a factor $1/(1 + m/M)$. Here M is the mass of the nucleus.

To handle this situation Bohr modified his second postulate, Eq. (5.4), to require

$$\mu v r = N, \quad N = 1, 2, 3, \ldots \tag{5.15}$$

Using μ instead of m in this equation takes into account the angular momentum of the nucleus as well as that of the electron. All the equations are identical with those derived before, except that the electron mass m is replaced by the reduced mass μ. In particular, the formula for the frequencies of the spectral lines becomes

$$\nu = R_M Z^2 \left(\frac{1}{N_f^2} - \frac{1}{N_i^2} \right),\tag{5.16}$$

where

$$R_M = \frac{\mu}{m} R_\infty.\tag{5.17}$$

The quantity R_M is the Rydberg constant for a nucleus of mass M. For the most extreme case of hydrogen R_M is less than R_∞ by about one part in 2000. If we evaluate R_H from Eq. (5.17) we find that the Bohr model, corrected for the finite nuclear mass, agrees with the experimental data to within three parts in 100,000

Atoms that follow the same pattern as hydrogen, but with different values of the parameters, are called *hydrogenic atoms*. They include systems in which an electron or the proton are replaced by an *exotic* particle (see below). They also include *Rydberg atoms*, an ordinary atom in which one electron has been elevated to a very high quantum state, and highly ionized atoms; that is, atoms whose electrons have been stripped away, leaving only one electron in an orbit around the nucleus. In Rydberg atoms the methods developed here serve only as approximations, in which the interactions between the lower-lying electrons are neglected.

An example of an exotic atom is *positronium*, which consists of a positron and an electron, revolving about their common center of mass, which lies halfway between them. The reduced mass is $\mu = m/2$. The corresponding Rydberg constant is $R_M = R_\infty/2$. The frequencies of the emitted spectral lines of positronium are given by

$$\nu = \frac{R_\infty}{2} \left(\frac{1}{N_f^2} - \frac{1}{N_i^2} \right).\tag{5.18}$$

The frequencies of the emitted lines are then half those of a hydrogen atom (with infinitely heavy nucleus), Z being equal to one for positronium and for hydrogen. The radius of the ground state of positronium is

$$r_{\text{pos}} = \frac{1}{\alpha\mu} = 2a_0. \tag{5.19}$$

Hence, for any quantum state N the radius of the electron is twice as great in the positronium atom as in the hydrogen atom (with infinitely heavy nucleus).

A *muonic* atom contains a nucleus of charge $Z\alpha$ and a negative *muon*, μ^-, moving about it. The μ^- is an elementary particle with charge $-e$ and a mass that is about 207 times as large as an electron mass. Such an atom is formed when a proton, or some other nucleus, captures a μ^-. The reduced mass of the system is $\mu = 186m$. Then, from Eq. (5.6), with $N = Z = 1$, and $\mu = 186m$, we obtain

$$r_1 = \frac{1}{186m\alpha} = 2.8 \times 10^{-3}\,\mathring{A}. \tag{5.20}$$

Therefore the μ^- is much closer to the nuclear (proton) surface than is an electron in a hydrogen atom. It is this feature that makes muonic atoms interesting, information about nuclear properties being revealed from their study. The binding energy of a muonic atom with $Z = 1$ is calculated from Eq. (5.11), with $N = 1$ and $\mu = 186\,m$; the result is $E = -2530\,eV$. The wavelength of the first line in the Lyman series is

$$\nu = R_M \left(1 - \frac{1}{4}\right), \tag{5.21}$$

with $R_M = (\mu/m)R_\infty$, so $\lambda = 1/\nu = 6.5\,\mathring{A}$, so that the Lyman lines lie in the x-ray part of the spectrum. X-ray techniques are therefore necessary to study the spectrum of muonic atoms.

Ordinary hydrogen contains about one part in 6000 of *deuterium*, or heavy hydrogen. This is a hydrogen atom whose nucleus contains a proton *and* a neutron. The spectrum would be identical to that of hydrogen were it not for the correction of finite nuclear mass. For a normal hydrogen atom

$$R_H = \frac{109737\,cm^{-1}}{\left(1 + \frac{1}{1836}\right)} = 109678\,cm^{-1}. \tag{5.22}$$

For an atom containing deuterium

$$R_D = \frac{109737\,cm^{-1}}{\left(1 + \frac{1}{2\times1836}\right)} = 109707\,cm^{-1}. \tag{5.23}$$

Hence, R_D is a bit larger than R_H, so that the spectral lines of the deuterium atom are shifted to slightly lesser wavelengths compared to hydrogen. Indeed, deuterium was discovered in 1932 by H. C. Urey following the observation of these shifted spectral lines. By increasing the concentration of the heavy isotope above its normal value in a hydrogen discharge tube, one can enhance the intensity of the deuterium lines which, ordinarily, are difficult to detect. One then readily observes pairs of hydrogen lines; the shorter wavelength members of the pair correspond exactly to those predicted from R_D. The resolution is easily obtained, the H_α-line pair being separated by about 1.8 Å, for example, several thousand times greater than the minimal resolvable separation.

Notes on Chapter 5

Information on Niels Bohr and the Bohr model can be found in *Hydrogen: The Essential Element*, by John Rigden [Rig03]. Also in that book are further details on hydrogenic atoms. Further recommended reading is *The Infancy of Atomic Physics: Hercules in His Cradle*, by Alex Keller [Kel83]. The presentation here follows Eisberg and Resnick [ER74].

Chapter 6

Interpretation of the Quantum Rules

6.1 The Sommerfeld–Wilson Quantization Conditions

The success of the Bohr model, as measured by its agreement with experiment, was at the time certainly very striking; but it only accentuated the mysterious nature of the postulates on which the model was based. One of the biggest mysteries was the question of the relation between Bohr's quantization of the angular momentum of an electron moving in a circular orbit and Planck's quantization of the total energy of an entity, such as an electron, executing simple harmonic motion. In 1916 some light was shed upon this question by Wilson [Wil15], and by Sommerfeld [Som16], who enunciated a set of rules for the quantization of any physical system for which the coordinates are periodic functions of time. These rules included both the Planck and the Bohr quantization as special cases. They were also of considerable use in broadening the range of applicability of the quantum theory. These rules can be stated as follows:

For any physical system in which the coordinates are periodic functions of time, there exists a quantum condition for each coordinate. The quantum conditions are

$$\oint p_q dq = 2\pi N_q, \tag{6.1}$$

where q is one of the coordinates, p_q is the conjugate momentum associated with that coordinate, N_q is a quantum number that takes on integral values, and $\oint$ means that the integration is taken over one complete period of the coordinate q. The momentum conjugate to a coordinate is in the sense of Hamiltonian mechanics, namely $\{q, p\} = 1$, where $\{A, B\}$ is the Poisson bracket of the two quantities.

The meaning of these rules can best be illustrated in terms of specific examples. Consider a one-dimensional simple harmonic oscillator. Its total energy can be written, in terms of position and momentum, as

$$E = E_{\text{kin}} + \frac{kx^2}{2}, \tag{6.2}$$

or

$$\frac{p_x^2}{2mE} + \frac{x^2}{2E/k} = 1.$$ (6.3)

The quantization integral $\oint p_x dx$ is most easily evaluated, for the relation between p_x and x that is imposed by this equation, if we consider a geometric interpretation. The relation between p_x and x is the equation of an ellipse. Any instantaneous state of motion of the oscillator is represented by some point in a plot of this equation on a two-dimensional space having coordinates p_x and x. Such a space (the $p - q$ plane) is the *phase space*, and the plot is a *phase diagram* of the linear oscillator. During one cycle of oscillation the point representing the position and the momentum of the particle travels once around the ellipse. The semiaxes a and b of the ellipse $p_x^2/b^2 + x^2/a^2 = 1$ are seen, by comparison with Eq. (6.3), to be

$$b = \sqrt{2mE} \quad \text{and} \quad a = \sqrt{2E/k}.$$ (6.4)

Now the area of an ellipse is πab. The value of the integral $\oint p_x dx$ is just equal to that area. Thus we obtain

$$\oint p_x dx = \pi ab.$$ (6.5)

In our case

$$\oint p_x dx = \frac{2\pi E}{\sqrt{k/m}},$$ (6.6)

but

$$\sqrt{k/m} = 2\pi\nu,$$ (6.7)

where ν is the frequency of the oscillation, so that

$$\oint p_x dx = \frac{E}{\nu}.$$ (6.8)

If we now use Eq. (6.1), the Sommerfeld–Wilson quantization rule, we have

$$\oint p_x dx = \frac{E}{\nu} = 2\pi N_x = 2\pi N,$$ (6.9)

or

$$E = 2\pi N\nu,$$ (6.10)

which is identical with Planck's quantization law.

Note that the allowed states of oscillation are represented by a series of ellipses in phase space, the area enclosed between successive ellipses always being $2\pi = 2\pi\hbar = h$. We find that the classical situation corresponds to $2\pi\hbar \to 0$, all values of E and hence all ellipses being allowed *if* that were true. The quantity $\oint p_x dx$ is called a *phase integral*; in classical physics it is the integral of the dynamical quantity called the *action* over one oscillation of the motion. Hence, the Planck energy quantization condition is equivalent to the *quantization of the action*.

The full quantum mechanical treatment of course gives $E = 2\pi\nu(N + \frac{1}{2})$, so that the Sommerfeld–Wilson quantization condition does not give the zero-point energy correctly.

We can also deduce the Bohr quantization of angular momentum from the Sommerfeld–Wilson rule, Eq. (6.1). An electron moving in a circular orbit of radius r has an angular momentum, $mvr = L$, which is constant. The angular coordinate is θ, which is a periodic function of time. That is, θ versus t is a sawtooth function, increasing linearly from zero to 2π rad in one period and repeating this pattern in each succeeding period. The quantization rule

$$\oint p_q dq = 2\pi N_q, \tag{6.11}$$

becomes, in this case,

$$\oint L d\theta = 2\pi N, \tag{6.12}$$

and

$$\oint L d\theta = L \int_0^{2\pi} d\theta = 2\pi L, \tag{6.13}$$

so that $L = N$, which is identical with Bohr's quantization law (5.2).

6.2 de Broglie's Wave Interpretation

In his doctoral thesis, presented in 1924 to the Faculty of Science at the University of Paris, de Broglie [Bro24] proposed the existence of matter waves. The thoroughness and originality of his thesis was recognized at once, but because of the apparent lack of experimental evidence, de Broglie's ideas were not considered to have any physical reality. It was Albert Einstein who recognized their importance and validity and in turn called them to the attention of other physicists. Five years later de Broglie won the Nobel Prize in physics, his ideas having been dramatically confirmed by experiment.

The hypothesis of de Broglie [Bro24] was that the dual, that is the wave-particle, behavior of radiation applies equally well to matter. Just as a photon has a light wave associated with it that governs its motion, so a material particle (e.g., an electron) has an associated matter wave that governs its motion. Since the universe is composed entirely of matter and radiation, de Broglie's suggestion is essentially a statement about a grand symmetry of nature. Indeed, he proposed that the wave aspects of matter are related to its particle aspects in exactly the same quantitative way that is the case for radiation. According to de Broglie, for matter *and* radiation alike the total energy of an entity is related to the frequency ν of the wave associated with its motion by the equation

$$E = 2\pi\nu, \tag{6.14}$$

and the momentum p of the entity is related to the wavelength λ of the associated wave by the equation

$$p = 2\pi/\lambda. \tag{6.15}$$

Here the particle concepts, energy E and momentum p, are connected through Planck's constant $h = 2\pi\hbar = 2\pi$, to the wave concepts, frequency ν and wavelength λ. Equation (6.15), in the following form, is called *de Broglie's relation*:

$$\lambda = \frac{2\pi}{p}. \tag{6.16}$$

It predicts the *de Broglie wavelength* λ of a *matter wave* associated with the motion of a material particle having a momentum p.

The wave nature of light propagation is not revealed in experiments in geometrical optics, for the important dimensions of the apparatus used there are very large compared to the wavelength of light. Geometrical optics involves ray propagation, which is similar to the trajectory motion of classical particles. When the characteristic dimension of an optical apparatus becomes comparable to, or smaller than, the wavelength of the light going through it, we are in the domain of physical optics, where diffraction effects are observed, and the wave nature of light propagation becomes apparent.

To observe wavelike aspects in the motion of matter, therefore, we need systems with apertures or obstacles of suitable small dimensions. Using apparatus with characteristic dimensions of 1 Å, wavelike aspects of the motion of electrons should be apparent.

Elsasser pointed out, in 1926, that the wave nature of matter might be tested in the same way that the wave nature of x-rays was first tested, namely by allowing a beam of electrons of appropriate energy to fall on a crystalline solid. The atoms of the crystal serve as a three-dimensional array of diffracting centers for the electron wave, and so they should strongly scatter electrons in certain characteristic directions, just as for x-ray diffraction. This idea was confirmed in experiments by Davisson and Germer in the United States and by G. P. Thomson in Scotland.

Going back to 1924, de Broglie applied his concepts to an electron orbiting a nucleus. The Bohr quantization of angular momentum can be written as in Eq. (5.4) as

$$mvr = pr = N, \quad N = 1, 2, 3, \ldots \tag{6.17}$$

where p is the momentum of an electron in an allowed orbit of radius r. If we substitute into this equation the expression for p in terms of the corresponding de Broglie wavelength,

$$p = \frac{2\pi}{\lambda}, \tag{6.18}$$

the Bohr equation becomes

$$\frac{2\pi r}{\lambda} = N, \tag{6.19}$$

or

$$2\pi r = N\lambda, \quad N = 1, 2, 3, \ldots \tag{6.20}$$

Thus *the allowed orbits are those in which the circumference of the orbit can contain exactly an integral number of de Broglie wavelengths.*

Imagine the electron moving in a circular orbit at constant speed. It has a wave associated with its motion. The wave, of wavelength λ, is wrapped repeatedly around the circular orbit. The resultant wave that is produced will have zero intensity at any point unless the wave at each traversal is exactly in phase at that point with the wave in other traversals. If the wave in each traversal is exactly in phase, they join on perfectly in orbits that accommodate integral numbers of de Broglie wavelengths, as illustrated in Figure 6.1. But the condition that this happens is just the condition that Eq. (6.20) be satisfied.

This wave picture gives no suggestion of progressive motion. Rather, it suggests standing waves, as in a stretched string of a given length. In a stretched string only certain wavelengths, or frequencies of vibration, are permitted. Once such modes are excited, the vibration goes on indefinitely if there is no damping. To get standing waves, however, we need oppositely directed traveling waves of equal amplitude. For the atom this requirement is satisfied by the fact that the electron can traverse an orbit in either direction and still have the same magnitude of angular momentum required by Bohr. The de Broglie standing wave

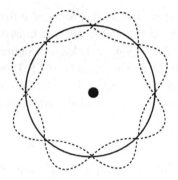

Figure 6.1.   de Broglie waves associated with an electron.

interpretation, illustrated in Figure 6.1, therefore provides a satisfying basis for Bohr's quantization rule, and, for this case, of the more general Sommerfeld–Wilson rule. The time-independent features of the standing wave associated with the electron in a stationary orbit of an atom make it possible to understand in a simple way why the electron does not emit electromagnetic radiation and spiral into the nucleus.

Notes on Chapter 6

The presentation here follows Eisberg and Resnick [ER74].

Chapter 7
Sommerfeld's Model for Non-Relativistic Electrons

One of the most important applications of the Sommerfeld–Wilson quantization rules is to the case of a hydrogen atom in which the electron can move, not only in *circular* orbits, but in *elliptical* orbits. As we have seen, this is the motion allowed in general in a $1/r$-potential by classical mechanics. In this chapter we deal with the application of this idea to slowly moving electrons, leaving its extension to relativistic particles to Chapter 12.

7.1 Assumptions of the Model

In non-relativistic physics the Hamiltonian for an attractive Coulomb potential is

$$H = \frac{\mathbf{p}^2}{2m} - \frac{Z\alpha}{r}, \tag{7.1}$$

where Z is the nuclear charge and α the fine-structure constant $\alpha \approx 1/137$. This equation can be written as

$$H + \frac{Z\alpha}{r} = E_{\text{kin}}. \tag{7.2}$$

From the identity

$$L^2 = (\mathbf{r} \times \mathbf{p})^2 = r^2 p^2 - (\mathbf{r} \cdot \mathbf{p})^2 \tag{7.3}$$

we infer, with $p_r = \mathbf{p} \cdot \hat{\mathbf{r}}$,

$$\mathbf{p}^2 = p_r^2 + \frac{L^2}{r^2} = p_r^2 + \frac{p_\theta^2}{r^2}, \tag{7.4}$$

where p_r is the momentum conjugate to r and $p_\theta = |\mathbf{L}|$ is the momentum conjugate to θ.

We calculate the radial momentum as

$$p_r = m\dot{r} = m\dot{\theta}\left(\frac{dr}{d\theta}\right) = \frac{p_\theta}{r^2}\left(\frac{dr}{d\theta}\right), \tag{7.5}$$

using $p_\theta = L = m\dot{\theta}r^2$.

For convenience, define the new variable

$$s = \frac{1}{r},$$ (7.6)

and note that

$$\frac{p_r}{p_\theta} = \frac{ds}{d\theta} = \frac{1}{r^2}\frac{dr}{d\theta} = \frac{1}{r^2}\left(\frac{\dot{r}}{\dot{\theta}}\right).$$ (7.7)

Introducing these variables into Eq. (7.1), we find that

$$H + Z\alpha s = \frac{p_\theta^2}{2m}\left(\left(\frac{ds}{d\theta}\right)^2 + s^2\right).$$ (7.8)

One differentiates Eq. (7.8) with respect to θ and obtains the linear differential equation:

$$\frac{d^2 s}{d\theta^2} + s - C = 0,$$ (7.9)

where we have used the abbreviation

$$C = \frac{Z\alpha m}{p_\theta^2}.$$ (7.10)

The general solution for Kepler orbits is

$$s = \frac{1}{r} = A\cos\theta + B\sin\theta + C,$$ (7.11)

where A and B are the integration constants and C is given in Eq. (7.10). Defining the distance of closest approach (perihelion) to occur at $\theta = 0$ makes $B = 0$, yielding the simpler equation:

$$\frac{1}{r} = A\cos\theta + C.$$ (7.12)

This result shows that the Kepler orbits have the form of conic sections.

Let us now express the elliptic orbits in standard geometrical terms, using the eccentricity e. If a is the semimajor axis of the ellipse, $a = c/(1 - e^2)$, then the orbit equation for the ellipse reads

$$\frac{1}{r} = \frac{1 + e\cos\theta}{c} = \frac{1}{a}\left(\frac{1 + e\cos\theta}{1 - e^2}\right).$$ (7.13)

Quantization is achieved by applying the Sommerfeld–Wilson quantization rules to the phase integrals for p_θ and p_r:

$$\oint p_\theta d\theta = \int_{\theta=0}^{\theta=2\pi} L \, d\theta = 2\pi n_\theta, \tag{7.14}$$

which yields $p_\theta = n_\theta$, and

$$\oint p_r \, dr = 2\pi n_r. \tag{7.15}$$

To evaluate the integral in Eq. (7.15) we use the expression Eq. (7.5) for p_r and obtain

$$p_r dr = \frac{p_r}{r^2} \left(\frac{dr}{d\theta}\right)^2 d\theta = p_\theta \frac{e^2 \sin^2 \theta}{(1+e\cos\theta)^2} d\theta. \tag{7.16}$$

Thus, the radial quantization condition, Eq. (7.15), takes the form

$$\frac{1}{2\pi} \int_{\theta=0}^{\theta=2\pi} \frac{e^2 \sin^2\theta \, d\theta}{(1+e\cos\theta)^2} = \frac{n_r}{n_\theta}. \tag{7.17}$$

The integral in this equation was evaluated by Sommerfeld [Som16]

$$\frac{1}{2\pi} \int_{\theta=0}^{\theta=2\pi} \frac{e^2 \sin^2\theta \, d\theta}{(1+e\cos\theta)^2} = (1-e^2)^{-\frac{1}{2}} - 1, \tag{7.18}$$

so

$$(1-e^2)^{-\frac{1}{2}} = 1 + \frac{n_r}{n_\theta}. \tag{7.19}$$

7.2 Results of the Model for Non-Relativistic Hydrogen Atoms

It is now straightforward to obtain Bohr's formula for the energy levels from these results. The two parameters of the orbit equation (a and e in Eq. (7.13)) are determined from the quantization conditions:

1. From Eq. (7.19) we know the eccentricity of the orbit:

$$\frac{1}{1-e^2} = \left(1 + \frac{n_r}{n_\theta}\right)^2. \tag{7.20}$$

2. From Eq. (7.10) we get the semimajor axis:

$$C = \frac{Z\alpha m}{n_\theta^2} = \frac{1}{a(1-e^2)},$$ (7.21)

and

$$A = eC = \frac{e}{a(1-e^2)}.$$ (7.22)

We then determine H from the energy equation, Eq. (7.8), and knowledge of the orbit $s = s(\theta)$. Eq. (7.8) becomes

$$H + Z\alpha(A\cos\theta + C) = \frac{n_\theta^2}{2m}(A^2\sin^2\theta + A^2\cos^2\theta + C^2 + 2AC\cos\theta)$$

$$= \frac{n_\theta^2}{2m}(A^2 + C^2 + 2AC\cos\theta).$$ (7.23)

Comparing the coefficients of $\cos\theta$ yields

$$Z\alpha = \frac{n_\theta^2}{m}C,$$ (7.24)

which holds because of Eq. (7.10). The constant term yields

$$H + Z\alpha C = \frac{n_\theta^2}{2m}(A^2 + C^2) = \frac{n_\theta^2}{2m}(e^2 + 1)C^2.$$ (7.25)

This leads to

$$H = \frac{n_\theta^2}{2m}\left(2 - (1 - e^2)\right)C^2 - Z\alpha C = -\left(\frac{m}{2}\right)\frac{(Z\alpha)^2}{(n_\theta + n_r)^2}.$$ (7.26)

For a given orbit the energy is constant

$$H = E = -\left(\frac{m}{2}\right)\frac{(Z\alpha)^2}{(n_\theta + n_r)^2}.$$ (7.27)

For the circular orbits $e = 0$, and from Eq. (7.20) it follows that $n_r = 0$. Thus, for Bohr's circular orbits

$$E = -\left(\frac{m}{2}\right)\frac{(Z\alpha)^2}{n_\theta^2}.$$ (7.28)

Defining $N = n_\theta + n_r$ we find that there are multiple orbits for $N > 1$, all of the same energy:

$$E_N = -\frac{m}{2}\frac{(Z\alpha)^2}{N^2}.$$ (7.29)

Since $n_\theta = 1, 2, 3, \ldots$ and $n_r = 0, 1, 2, 3, \ldots$, N can take on the values

$$N = 1, 2, 3, 4, \ldots \tag{7.30}$$

For a given value of N, n_θ can assume only the values

$$n_\theta = 1, 2, 3, \ldots, N. \tag{7.31}$$

The integer N is called the *principal quantum number*, and n_θ is called the *azimuthal quantum number*. $n_\theta = 0$ is not allowed; for $n_r = 0$ it would correspond to an orbit going through the force center.

7.3 The Eccentricity

We can write Eq. (7.19) in the form

$$e = \sqrt{1 - \left(\frac{n_\theta}{N}\right)^2}. \tag{7.32}$$

This shows that the shape of the orbit (the ratio of the semiminor to the semimajor axes) is determined by the ratio of n_θ to N. For $n_\theta = N$ the orbits are circular. Figure 7.1 shows possible orbits corresponding to the first three values of the principle quantum number. Corresponding to each value of the principal quantum number N there are N different allowed orbits. One of these, the circular orbit, is just the orbit described by the original Bohr model. The others are elliptical. But despite the very different paths followed by an electron moving in the different possible orbits for a given N, Eq. (7.29) tells us that the total energy of the electron is the same. The total energy of an electron depends only on N. The several orbits characterized by a common value of N are said to be *degenerate*. The energies of different states of motion "degenerate" to the same total energy.

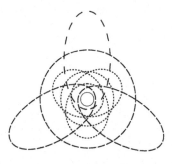

Figure 7.1. The Sommerfeld model of the hydrogen atom.

This degeneracy in the total energy of an electron, due to orbits of very different shape but common N, is the result of a very delicate balance between potential and kinetic energy, which is characteristic of treating the inverse square Coulomb force by the methods of classical mechanics. Exactly the same phenomena is found in planetary or satellite motion, which is governed by the inverse square gravitational force. For instance, a satellite may be launched into any one of a whole family of elliptical orbits, all of which correspond to the same state of the same total energy and have the same semimajor axis. Of course, there is no quantization of the orbit parameters in these macroscopic cases, but as far as degeneracy is concerned they are completely analogous to the case of the hydrogen atom.

To get the complete degeneracy of the electron, neglecting the spin, we turn to a group-theoretical analysis in the next chapter.

Notes on Chapter 7

The history of atomic models is told in Keller [Kel83]. The importance of hydrogen in this context is emphasized by Rigden [Rig03]. This chapter develops a non-relativistic version of Sommerfeld's work [Som16], following Biedenharn [Bie83]. See also Sommerfeld's book [Som24].

Chapter 8
Quantum Mechanics of Hydrogenic Atoms

Heisenberg [Hei25], on the basis of the spectral data of hydrogen, wrote a paper in which he formulated for the first time the canonical commutation relations between the basic variables of position and momenta, $[x_i, p_j] = i\delta_{ij}$. He gave his manuscript to Born, who submitted it to the Zeitschrift für Physik. Born recognized that Heisenberg's equations could be expressed in matrix form. The formulation of the matrix formalism of quantum mechanics was published in two papers in 1925, by Born and Jordan [BJ25], and by Born, Heisenberg and Jordan [MBJ26].

The immediate question was whether the new theory was capable of explaining the spectrum of hydrogen, as the Bohr model had done. The answer was given by Pauli [Pau26], and independently by Dirac [Dir26], towards the end of 1925. Pauli obtained the same formula that Bohr had obtained in 1913, only this time the route to the formula was a coherent theory — the new theory of quantum mechanics.

In the following we shall show, following Pauli, that the basic commutation relations, and a general knowledge of the symmetry group SO(4), lead to the Balmer formula, and a first estimate of the degeneracy of the energy levels.

The classical Kepler problem is characterized by six conserved quantities: the three components of the orbital angular momentum $\mathbf{L} = (L_1, L_2, L_3)$ and the three components of the Laplace vector, $\mathcal{A} = (\mathcal{A}_1, \mathcal{A}_2, \mathcal{A}_3)$, satisfying the two constraints $\mathcal{A} \cdot \mathbf{L} = 0$ and $|\mathcal{A}|^2 = 2EL^2/((Z\alpha)^2 m) + 1$. We shall see that these relations all have their counterparts in the quantum mechanical case.

8.1 Quantization

In general, we go from a classical to a quantum system by following the *Dirac quantization procedure* [Dir53], i.e., we replace observable quantities by Hermitian operators, and Poisson brackets of observables by commutators, according to the rule $\{,\} \rightarrow i[,]$.

For the basic observables, position x_i and momentum p_j, this means that $\{x_i, p_j\} = \delta_{ij}$ is replaced by $[x_i, p_j] = i\delta_{ij}$. More complex commutators are reduced by analogs of the rules (a)–(d) of Section 3.3, e.g.,

$$[A, BC] = [A, B]C + B[A, C], \qquad (8.1)$$

which follows from

$$[A, BC] = ABC - BCA = ABC - BAC + BAC - BCA = [A, B]C + B[A, C]. \qquad (8.2)$$

For an electron in a Coulomb field the Hamiltonian is

$$H = \frac{\mathbf{p}^2}{2m} - \frac{Z\alpha}{r}. \qquad (8.3)$$

The components of the angular momentum are

$$L_i = \epsilon_{ijk} x_j p_k. \qquad (8.4)$$

These are Hermitian:

$$L_i = \epsilon_{ijk} x_j p_k, \quad L_i^\dagger = \epsilon_{ijk} p_k x_j = \epsilon_{ijk} x_j p_k = L_i. \quad i = 1, 2, 3. \qquad (8.5)$$

The commutator of the Hamiltonian with a component of the angular momentum is

$$[H, L_i] = \frac{1}{2m}[\mathbf{p}^2, L_i] = \frac{1}{2m}\epsilon_{ijk}[\mathbf{p}^2, x_j p_k]$$

$$= \frac{1}{2m}\epsilon_{ijk}\left(p_l[p_l, x_j]p_k + [p_l, x_j]p_k p_l\right)$$

$$= \frac{-i}{2m}\epsilon_{ijk}(p_j p_k + p_k p_j) = 0. \qquad (8.6)$$

The angular momentum commutes of course with arbitrary functions of r. This tells us that $\mathbf{L}$ is a constant vector.

If we try to take the Laplace vector $\mathcal{A} = (1/2m)(\mathbf{p} \times \mathbf{L}) - \hat{r}$ over to quantum mechanics it won't work, because $(\epsilon_{ijk} p_j L_k)^\dagger = \epsilon_{ijk} L_k p_j = -\epsilon_{ikj} L_k p_j$, or

$$(\mathbf{p} \times \mathbf{L})^\dagger = -\mathbf{L} \times \mathbf{p}, \qquad (8.7)$$

is not Hermitian. This may be easily corrected,

$$\mathcal{A} = \frac{1}{2Z\alpha m}(\mathbf{p} \times \mathbf{L} - \mathbf{L} \times \mathbf{p}) - \hat{r} \qquad (8.8)$$

is Hermitian, and reduces to the Laplace vector in the classical limit. More formally, we may say that quantizing the classical vector $\mathcal{A}$ according to the Weyl–Moyal quantization prescription [Tak08]

$$\mathcal{A}_i = \frac{1}{2Z\alpha m}\epsilon_{ijk}\{p_j, L_k\} - x_i/r \tag{8.9}$$

leads to Eq. (8.8). Now and in the following the brackets $\{,\}$ are taken to mean the *anticommutator*: $\{A, B\} = AB + BA$.

To show that $\mathcal{A}$ is a constant of the motion we wish to prove $[H, \mathcal{A}] = 0$. We do this by repeating the calculation of Section 3.4, paying attention this time to the ordering of factors, and using the basic commutation relation $[x_i, p_j] = i\delta_{ij}$ instead of the Poisson bracket $\{x_i, p_j\} = \delta_{ij}$. Equation (3.64) is now

$$(Z\alpha m)[H, \mathcal{A}_i] = \frac{\epsilon_{ijk}}{2}\{[H, p_j], L_k\} - \frac{Z\alpha}{2}\{p_j, [p_j, x_i/r]\}$$

$$= \frac{iZ\alpha\epsilon_{ijk}}{2}\{x_j/r^3, L_k\} + \frac{iZ\alpha}{2}\{p_j, \delta_{ij}/r - x_ix_j/r^3\}, \tag{8.10}$$

using the rule $[p_i, g] = -i\partial g/\partial x_i$, for g a function of x_i, and the first term is

$$\frac{iZ\alpha\epsilon_{ijk}\epsilon_{kls}}{2}\left(\frac{x_j}{r^3}x_lp_s + p_sx_l\frac{x_j}{r^3}\right) = \frac{iZ\alpha}{2}(\delta_{il}\delta_{js} - \delta_{is}\delta_{jl})\left\{p_s, \frac{x_jx_l}{r^3}\right\}, \tag{8.11}$$

which cancels the second term.

The quantum mechanical generalization of Eq. (2.38) is

$$\mathcal{A}\cdot\mathbf{L} = \mathbf{L}\cdot\mathcal{A} = 0. \tag{8.12}$$

We prove this as follows: Use $\mathbf{L} \times \mathbf{p} = \mathbf{p} \times \mathbf{L} - 2i\mathbf{p}$ to write

$$\mathcal{A} = \frac{1}{Z\alpha m}(\mathbf{p} \times \mathbf{L} - i\mathbf{p}) - \hat{r}. \tag{8.13}$$

Then, since $\mathbf{p}\cdot\mathbf{L} = \mathbf{r}\cdot\mathbf{L} = 0$,

$$(Z\alpha m)\mathcal{A}\cdot\mathbf{L} = \epsilon_{ijk}p_jL_kL_i = \frac{1}{2}\epsilon_{ijk}p_j[L_k, L_i] = i\mathbf{p}\cdot\mathbf{L} = 0. \tag{8.14}$$

Similarly, $\mathbf{L}\cdot\mathcal{A} = 0$.

8.2 Quantum Mechanical Relation Between $|\mathcal{A}|^2$ and $\mathbf{L}^2$

Finally, we wish to prove the generalization of Eq. (2.40):

$$|\mathcal{A}|^2 = \frac{2H}{(Z\alpha)^2 m}(L^2 + 1) + 1. \tag{8.15}$$

We shall use the quantum identities

$$(a) \quad (\mathbf{p} \times \mathbf{L}) \cdot (\mathbf{p} \times \mathbf{L}) = p^2 L^2,$$

$$(b) \quad (\mathbf{p} \times \mathbf{L}) \cdot \mathbf{p} = 2ip^2,$$

$$(c) \quad \mathbf{p} \cdot (\mathbf{p} \times \mathbf{L}) = 0,$$

$$(d) \quad (\mathbf{p} \times \mathbf{L}) \cdot \mathbf{r} = L^2 + 2i(\mathbf{p} \cdot \mathbf{r}),$$

$$(e) \quad \mathbf{r} \cdot (\mathbf{p} \times \mathbf{L}) = L^2. \tag{8.16}$$

Exercise 8.1 Prove these identities.

We calculate

$$(Z\alpha m)^2(|\mathcal{A}|^2 - 1) = (\mathbf{p} \times \mathbf{L} - i\mathbf{p})^2 - (Z\alpha m)(\mathbf{p} \times \mathbf{L} - i\mathbf{p}) \cdot \hat{r}$$

$$- (Z\alpha m)\hat{r} \cdot (\mathbf{p} \times \mathbf{L} - i\mathbf{p}). \tag{8.17}$$

Use the identities (8.16) to get

$$(\mathbf{p} \times \mathbf{L} - i\mathbf{p})^2 = \mathbf{p}^2(L^2 + 1),$$

$$(\mathbf{p} \times \mathbf{L} - i\mathbf{p}) \cdot \hat{r} = \frac{L^2}{r} + i\mathbf{p} \cdot r \frac{1}{r}, \tag{8.18}$$

$$\hat{r} \cdot (\mathbf{p} \times \mathbf{L} - i\mathbf{p}) = \frac{L^2}{r} - \frac{i}{r} r \cdot \mathbf{p}.$$

We finally obtain

$$|\mathcal{A}|^2 - 1 = (Z\alpha m)^{-2} \left(\mathbf{p}^2(L^2 + 1) - \frac{2Z\alpha m}{r} L^2 - \frac{2Z\alpha m}{r} \right)$$

$$= \frac{2}{(Z\alpha)^2 m} \left(\frac{\mathbf{p}^2}{2m} - \frac{Z\alpha}{r} \right) (L^2 + 1) = \frac{2H}{(Z\alpha)^2 m}(L^2 + 1). \tag{8.19}$$

Thus we see that in the quantum mechanical case we have, as in the classical case, six constants of the motion, $\mathbf{L} = (L_1, L_2, L_3)$ and $\mathcal{A} = (\mathcal{A}_1, \mathcal{A}_2, \mathcal{A}_3)$, which satisfy the two constraints, Eq. (8.12) and Eq. (8.15).

8.3 Pauli's Hydrogenic Realization of so(4)

In the quantum mechanical case the commutators between the components of $\mathbf{L}$ and the components of $\mathcal{A}$ are of interest. We shall see that, properly normalized, these commutators are those of an so(4)-algebra.

We can calculate $[L_i, L_j]$ from

$$[L_i, L_j] = i\epsilon_{jmn}(x_m[L_i, p_n] + [L_i, x_m]p_n). \tag{8.20}$$

The individual commutators follow from the definition

$$L_i = \epsilon_{imn}x_m p_n, \tag{8.21}$$

$$[L_i, x_m] = \epsilon_{irs}[x_r p_s, x_m] = \epsilon_{irs}x_r[p_s, x_m] = -i\epsilon_{irm}x_r = i\epsilon_{imr}x_r, \tag{8.22}$$

and similarly

$$[L_i, p_m] = i\epsilon_{qim}p_q, \tag{8.23}$$

so

$$[L_i, L_j] = i\epsilon_{jnm}(\epsilon_{inq}x_q p_m + \epsilon_{imq}x_n p_q) = i(x_i p_j - x_j p_i) = i\epsilon_{ijk}L_k. \tag{8.24}$$

In general, we say that a vector operator **v**, whose components satisfy with the components of the angular momentum

$$[L_i, v_j] = i\epsilon_{ijk}v_k, \tag{8.25}$$

transforms as a *vector operator*.

The commutators $[L_i, \mathcal{A}_j] = i\epsilon_{ijk}\mathcal{A}_k$ are the statement that $\mathcal{A}$ transforms as a vector operator with respect to **L**; this is obvious, since $\mathcal{A}$ is made up of such vectors.

We now have to calculate $[\mathcal{A}_i, \mathcal{A}_j]$. The calculation consists of a repetition of that with Poisson brackets, this time taking into account the ordering of factors. We also have to take into account the additional terms in the quantum mechanical expression for $\mathcal{A}$, Eq. (8.8), compared to the classical expression, Eq. (2.22). Now we get

$$(Z\alpha m)[p_i, \mathcal{A}_j] = \frac{1}{2}\epsilon_{jsn}[p_i, p_s L_n] - \frac{1}{2}[p_i, L_s p_n] - Z\alpha m[p_i, x_j/r]$$
$$= i(p_i p_j - \mathbf{p}^2\delta_{ij}) + i(Z\alpha m)(\delta_{ij}/r - x_i x_j/r^3). \tag{8.26}$$

To calculate $[p_i, 1/r]$ we use the analog of the rule (3.46), generalized to the quantum context, $[p_i, g] = -i\partial g/\partial x_i$, for $g(x)$ a function of x.

$$(Z\alpha m)[x_i/r, \mathcal{A}_j] = \epsilon_{jsn}([x_i/r, p_s L_n] - [x_i/r, L_s p_n])$$

$$= i\left(\left(\epsilon_{ijk}\frac{L_k}{r} - \frac{1}{2r^3}\epsilon_{jsn}(x_i x_s + L_n + L_n x_i x_s) \right. \right.$$

$$\left. \left. + \frac{1}{2}\left(p_i\frac{x_j}{r} + \frac{x_j}{r}p_i\right) - \frac{1}{2}\delta_{ij}\left(p_s\frac{x_s}{r} + \frac{x_s}{r}p_s\right) \right). \quad (8.27)$$

The latter yields

$$(Z\alpha m)^2 \epsilon_{ijk}[x_i/r, \mathcal{A}_j] = -2imL_k/r^3, \quad (8.28)$$

which is the analog of Eq. (3.72).

The main calculation is

$$(Z\alpha m)^2 \epsilon_{ijk}[\mathcal{A}_i, \mathcal{A}_j]$$

$$= (Z\alpha m)\epsilon_{ijk}\left(\frac{1}{2}\epsilon_{isn}([p_s L_n, \mathcal{A}_j] + [L_n p_s, \mathcal{A}_j]) - (Z\alpha m)[x_i/r, \mathcal{A}_j]\right)$$

$$= (Z\alpha m)\epsilon_{ijk}\left(\frac{1}{2}\epsilon_{isn}(p_s[L_n, \mathcal{A}_j] + [p_s, \mathcal{A}_j]L_n \right.$$

$$\left. + L_n[p_s, \mathcal{A}_j] + [L_n, \mathcal{A}_j]p_s) - (Z\alpha m)[x_i/r, \mathcal{A}_j]\right). \quad (8.29)$$

We evaluate the terms separately. The first two expressions are identical to those in the classical evaluation:

$$(Z\alpha m)\epsilon_{ijk}\epsilon_{isn}p_s[L_n, \mathcal{A}_j] = i(-\mathbf{p}^2 + (Z\alpha m)/r)L_k. \quad (8.30)$$

$$(Z\alpha m)\epsilon_{ijk}\epsilon_{isn}[p_s, \mathcal{A}_j]L_n = i(-\mathbf{p}^2 + (Z\alpha m)/r)L_k. \quad (8.31)$$

The next two expressions are new:

$$(Z\alpha m)\epsilon_{ijk}\epsilon_{isn}L_n[p_s, \mathcal{A}_j]$$

$$= i\epsilon_{ijk}\epsilon_{isn}L_n(p_s p_j - \mathbf{p}^2\delta_{sj} + (Z\alpha m)\delta_{sj}/r - (Z\alpha m)x_s x_j/r^3)$$

$$= i(\delta_{js}\delta_{kn} - \delta_{jn}\delta_{ks})L_n(p_s p_j - \mathbf{p}^2\delta_{sj} + (Z\alpha m)\delta_{sj}/r - (Z\alpha m)x_s x_j/r^3)$$

$$= iL_k(-\mathbf{p}^2 + (Z\alpha m)/r). \quad (8.32)$$

$$(Z\alpha m)\epsilon_{ijk}\epsilon_{isn}[L_n, \mathcal{A}_j]p_s = i(Z\alpha m)\epsilon_{ijk}\epsilon_{isn}\epsilon_{njq}\mathcal{A}_q p_s$$

$$= i(Z\alpha m)(\delta_{js}\delta_{kn} - \delta_{jn}\delta_{ks})\epsilon_{njq}\mathcal{A}_q p_s = i(Z\alpha m)\epsilon_{kjq}A_q p_j$$

$$= i\epsilon_{kjq}\left(\frac{1}{2}\epsilon_{qmn}(p_m L_n + L_n p_m) - (Z\alpha m)x_q/r\right)p_j$$

$$= \left(\frac{1}{2}(\delta_{km}\delta_{jn} - \delta_{kn}\delta_{jm})(p_m L_n + L_n p_m) - (Z\alpha m)\epsilon_{kjq}x_q/r\right)p_j$$

$$= iL_k(-\mathbf{p}^2 + (Z\alpha m)). \tag{8.33}$$

Combining these results, we get, as in the classical calculation involving Poisson brackets:

$$[\mathcal{A}_i, \mathcal{A}_j] = i\epsilon_{ijk}\left(\frac{-2H}{(Z\alpha)^2 m}\right)L_k = i\epsilon_{ijk}\mathcal{N}^{-2}L_k, \tag{8.34}$$

with

$$\mathcal{N}^{-2} = \frac{-2H}{(Z\alpha)^2 m}. \tag{8.35}$$

The appearance of the hydrogen Hamiltonian in (8.34) shows that $\mathcal{A}$ and $\mathbf{L}$ do not close under commutation to form a Lie algebra.

For the bound state problem H has nonzero negative eigenvalues, so that we may define $(\mathcal{N}^2)^{-\frac{1}{2}}$ and normalize the operator $\mathcal{A}$:

$$\mathbf{A} - (\mathcal{N}^2)^{\frac{1}{2}}\mathcal{A}. \tag{8.36}$$

Now the commutation relations become

$$[L_i, L_j] = i\epsilon_{ijk}L_k, \quad [L_i, A_J] = i\epsilon_{ijk}A_k, \quad [A_i, A_j] = i\epsilon_{ijk}L_k. \tag{8.37}$$

The $\mathbf{L}$ and $\mathbf{A}$ constitute the generators of an so(4) algebra, and provide a realization of the so(4) generators, called Pauli's hydrogenic realization. Alternatively we could consider continuum states and replace H with a positive energy and use (8.36) without the minus sign to obtain the so(3,1) commutation relations of the Lorentz group. Indeed, Dahl [Dah95] argues that the form of the Laplace vector is determined by the fact that it is simply related to the generator of Lorentz boosts.

The so(4) commutation relations can be transformed to a simpler canonical form. If we define the two vector operators

$$\mathbf{M} = \frac{1}{2}(\mathbf{L} + \mathbf{A}), \quad \mathbf{N} = \frac{1}{2}(\mathbf{L} - \mathbf{A}), \tag{8.38}$$

then the components of **M** and **N** satisfy the commutation relations

$$[M_i, M_j] = i\epsilon_{ijk}M_k, \quad [N_i, N_j] = i\epsilon_{ijk}N_k, \quad [M_i, N_j] = 0, \qquad (8.39)$$

which are equivalent to the original ones since (8.38) represents a real linear transformation of the original generators. The new commutation relations are simpler since **M** and **N** are two commuting angular momentum vectors. Therefore so(4) is the direct sum of so(3) $\oplus$ so(3) subalgebras. The first is generated by **M** and the second is generated by **N**.

The so(4) algebra has two Casimir operators; these are quadratic operators commuting with all the generators:

$$C_1 = \mathbf{L}^2 + \mathbf{A}^2, \quad C_2 = \frac{1}{2}(\mathbf{L}\cdot\mathbf{A} + \mathbf{A}\cdot\mathbf{L}). \qquad (8.40)$$

Exercise 8.2 Show that C_1 and C_2 are Casimir operators:

$$[C_1, \mathbf{L}] = [C_1, \mathbf{A}] = [C_2, \mathbf{L}] = [C_2, \mathbf{A}] = 0. \qquad (8.41)$$

8.4 so(4) and the Spectrum of Hydrogenic Atoms

Substituting (8.36) into (8.12) and (8.15) we obtain the important equations

$$\mathbf{A}\cdot\mathbf{L} = \mathbf{L}\cdot\mathbf{A} = 0, \quad |\mathbf{A}|^2 + \mathbf{L}^2 + 1 = \mathcal{N}^2. \qquad (8.42)$$

These give the values of the Casimir operators of so(4). The first tells us that $C_2 = 0$. The second equation contains the quadratic invariant $\mathbf{A}\cdot\mathbf{A} + \mathbf{L}\cdot\mathbf{L}$ and tells us that $C_1 = \mathcal{N}^2 - 1$. The irreducible representations of so(4) are given by the eigenvalues (C_1, C_2).

The equations for C_1 and C_2 may also be used to obtain the Bohr formula for the energy levels as follows. From $C_2 = 0$ it is clear that

$$M^2 = N^2 = \frac{1}{4}(L^2 + |\mathbf{A}|^2). \qquad (8.43)$$

Since the eigenvalues of M^2 and N^2 are $j_1(j_1 + 1)$ and $j_2(j_2 + 1)$, respectively, it follows that $j_1 = j_2$. Therefore, our realization of **A** using the modified Laplace vector does not give all the irreducible representations of so(4); only the so-called diagonal ones with $j_1 = j_2$.

This is analogous to the use of **L** as a realization of the so(3) generators. Only integral values of the angular momentum are realized. The values $j = 1/2, 3/2, \ldots$ are not possible with the realization $\mathbf{L} = \mathbf{r} \times \mathbf{p}$ in terms of coordinates and momenta.

The equation $C_1 = \mathcal{N}^2 - 1$ may be rewritten as

$$L^2 + |\mathbf{A}|^2 + 1 = \mathcal{N}^2, \qquad (8.44)$$

or substituting (8.43)

$$4M^2 + 1 = \mathcal{N}^2. \tag{8.45}$$

Replacing M^2 by its eigenvalue $j_1(j_1 + 1)$ we find

$$4j_1(j_1 + 1) + 1 = 4j_1^2 + 4j_1 + 1 = (2j_1 + 1)^2 = \mathcal{N}^2. \tag{8.46}$$

Since j_1 takes the values $j_1 = 0, 1/2, 1, 3/2, \ldots, \mathcal{N}^2$ takes the values $1, 4, 9, \ldots$ That is, the operator $\mathcal{N}^2$ takes the eigenvalues

$$\mathcal{N}^2 \to N^2. \tag{8.47}$$

Finally, using Eq. (8.35), we obtain the Bohr formula

$$E_N = -\frac{m}{2}\frac{(Z\alpha)^2}{N^2} \tag{8.48}$$

for the energy levels of the hydrogenic atom. We see that N is the principal quantum number. Given that $C_2 = 0$, the energy levels are classified by the single number C_1, or N.

We see that there is nothing "accidental" about the so-called "accidental degeneracy" of the energy levels. This term arose because the degeneracy of the $2l + 1$ hydrogenic states $|Nlm>$ for fixed N and l, corresponding to the energy E_N, could be explained since $[\mathbf{L}, H] = 0$, but the full degeneracy of the energy level,

$$\sum_{l=0}^{N-1}(2l + 1) = 2\frac{N(N-1)}{2} + N = N^2, \tag{8.49}$$

cannot be explained on the basis of SO(3) alone, since SO(3) does not guarantee that the different values of the angular momentum l all have the same energy. However, it is SO(4) that is the geometrical symmetry group, not SO(3). In fact it follows from $j_1 = j_2$ that the degeneracy of E_N is N^2, since there are $N^2 = (2j_1 + 1)^2$ basis vectors in the direct product space corresponding to the energy E_N.

It must be emphasized that N^2 does not give the *full* degeneracy of the levels, which is experimentally $2N^2$. The further factor of two comes from the *spin* of the electron, which has so far not entered our considerations. Another indication of this fact is discussed below.

Classically the eccentricity e is the length of the vector $\mathcal{A}$. We might attempt to take this relation over to quantum mechanics. From Eqs. (8.36), (8.42), and (8.47) we find that

$$|\mathcal{A}|^2 = \mathcal{A}\cdot\mathcal{A} = e^2 = 1 - \mathcal{N}^{-2}(L^2 + 1). \tag{8.50}$$

Thus for $L^2 \to l(l+1)$, $\mathcal{N}^2 \to N^2$, we find that the eccentricity becomes

$$e = \left(1 - \frac{l^2 + l + 1}{N^2}\right)^{\frac{1}{2}}. \tag{8.51}$$

For circular orbits, the eccentricity vanishes classically. Quantum mechanically, we shall see that the circular orbits (nodeless radial probability density) correspond to $l = N - 1$; e is then small ($\approx N^{-\frac{1}{2}}$ for N large), but *does not vanish* (except for $N = 1$).

This result for the eccentricity is strictly applicable to the orbits of spinless particles. For the electron, which has spin-$\frac{1}{2}$, it is not applicable. In Chapter 10 we shall find the value of the eccentricity for the orbits of the spinning electron.

Notes on Chapter 8

This chapter is a review of Pauli's paper [Pau26]. If you need to review the derivation of quantum mechanical commutation relations, or the application of group-theoretical symmetry considerations, an excellent place to look is the very lucid book by Adams [Ada94].

Chapter 9
The Schrödinger Equation
and the Confluent Hypergeometric Functions

In late 1925 Schrödinger read de Broglie's paper in which it was proposed that particles have an associated wavelength. Schrödinger went to work attempting to generalize de Broglie's concept of matter waves. The first sentence of Schrödinger's classic paper [Sch26] reads as follows: "In this paper I wish to consider, first, the simplest case of the hydrogen atom, and show that the customary quantum conditions can be replaced by another postulate, in which the concept of 'whole numbers', merely as such, is not introduced." In modern language Schrödinger is saying that he can dispense with the use of quantum conditions; they are replaced by the requirement that the argument of the hypergeometric function describing the solution of the wave equation must be a nonzero negative integer if the wave function is to fall off at infinity (see below).

The mathematics employed in the paper was familiar to physicists — it involved the solutions of differential equations. Schrödinger set up a wave equation in a form appropriate for the hydrogen atom — a negatively charged electron orbiting a positively charged nucleus. The solutions are normalizable only for discrete values of the energy. In 1926 Born suggested the interpretation of the solution of the wave equation: The magnitude of the quantity $|\psi|^2$, where ψ is the solution of the wave equation, is a measure of the probability density of the electron. Schrödinger's wave equation gives the maximum magnitude of $|\psi|^2$ in the ground state of the atom at a distance of 0.529 Å from the nucleus, in agreement with the Bohr radius, and the known size of the atom, whose diameter is about 1 Å.

9.1 The Schrödinger Equation and Its Solutions

The Schrödinger equation is obtained by replacing the classical Hamilton function by an operator, through the prescription $p = -i\nabla$, and applying it to a

The Supersymmetric Dirac Equation

wave function:

$$H\psi(\boldsymbol{r}) = \left[-\frac{1}{2m}\nabla^2 - \frac{Z\alpha}{r} \right]\psi(\boldsymbol{r}) = E\psi(\boldsymbol{r}). \qquad (9.1)$$

This equation is most naturally solved by transforming from Cartesian to spherical polar coordinates, in which system it is completely separable. We use the identity

$$L^2 = r^2\mathbf{p}^2 - \mathbf{r}(\mathbf{r}\cdot\mathbf{p})\cdot\mathbf{p}. \qquad (9.2)$$

The component of ∇ in the direction of $\mathbf{r}$ is $\partial/\partial r$, hence

$$\mathbf{r}\cdot\mathbf{p} = -ir\frac{\partial}{\partial r}, \qquad (9.3)$$

and

$$\mathbf{L}^2 = r^2\mathbf{p}^2 + r^2\frac{\partial^2}{\partial r^2} + 2r\frac{\partial}{\partial r}, \qquad (9.4)$$

or, more compactly

$$\mathbf{L}^2 = r^2\mathbf{p}^2 + \frac{\partial}{\partial r}\left(r^2\frac{\partial}{\partial r} \right). \qquad (9.5)$$

We can formulate this in terms of the momentem conjugate to r, which satisfies $[r, p_r] = i$. Then

$$p_r = -\frac{i}{r}\frac{\partial}{\partial r}r = -i\left(\frac{\partial}{\partial r} + \frac{1}{r} \right). \qquad (9.6)$$

Exercise 9.1 Show that p_r, as defined above, is canonically conjugate to r.

Exercise 9.2 Show that the Moyal–Weyl quantization prescription, applied to $\mathbf{p}$ and $\hat{\boldsymbol{r}}$, yields p_r.

Exercise 9.3 Show that p_r is Hermitian.

This yields

$$p_r^2 = -\frac{\partial^2}{\partial r^2} - \frac{2}{r}\frac{\partial}{\partial r} = -\frac{1}{r^2}\frac{\partial}{\partial r}r^2\frac{\partial}{\partial r}, \qquad (9.7)$$

and

$$\mathbf{p}^2 = p_r^2 + \frac{\mathbf{L}^2}{r^2}, \qquad (9.8)$$

as the quantum-mechanical counterpart to Eq. (7.4).

With this, solutions to Eq. (9.1) are found to be of the form

$$\psi(r, \theta, \phi) = R(r)Y(\theta, \phi), \tag{9.9}$$

where the $Y(\theta, \phi)$ are spherical harmonics and $R(r)$, the radial part of the wave function, satisfies

$$-2m(H - E) = -p_r^2 + \left[2mE - \frac{2Z\alpha m}{r} - \frac{l(l+1)}{r^2}\right] R$$

$$= \frac{1}{r^2}\frac{d}{dr}r^2\frac{dR}{dr} + \left[2mE + \frac{2Z\alpha m}{r} - \frac{l(l+1)}{r^2}\right] R = 0, \tag{9.10}$$

subject to the conditions that $\lim_{r \to 0} rR(r) = 0$ and $rR(r)$ be square integrable.

We introduce the dimensionless variable

$$x = 2\left(\frac{Z\alpha m}{\lambda}\right)r, \tag{9.11}$$

and obtain

$$x^2\frac{d^2R}{dx^2} + 2x\frac{dR}{dx} + [-\mu x^2 + \lambda x - l(l+1)]R = 0. \tag{9.12}$$

We have defined the parameter μ by

$$\mu = \frac{-E}{2m}\frac{\lambda^2}{(Z\alpha)^2}. \tag{9.13}$$

The parameter $\lambda > 0$ is yet to be determined. For bound states ($E < 0$) we also have $\mu > 0$.

Next we factor out the behavior near $x = 0$ and $x = \infty$, and look for solutions of the type

$$R(x) = x^l e^{-\sqrt{\mu}x}u(x). \tag{9.14}$$

For $u(x)$ we get the differential equation

$$xu''(x) + 2[(l+1) - \sqrt{\mu}x]u'(x) - [2(l+1)\sqrt{\mu} - \lambda]u(x) = 0. \tag{9.15}$$

With $\sqrt{\mu} = 1/2$ this becomes

$$xu''(x) + [2(l+1) - x]u'(x) - [l+1 - \lambda]u(x) = 0. \tag{9.16}$$

This is of the form of the confluent hypergeometric equation,

$$\left(x\frac{d^2}{dx^2} + (c - x)\frac{d}{dx} - a\right) {}_1F_1(a; c; x) = 0, \tag{9.17}$$

with $a = l+1-\lambda$ and $c = 2l+2$. The general solution of Eq. (9.17) is proportional to a confluent hypergeometric function,

$$u(x) = A_1 F_1(l + 1 - \lambda; 2l + 2; x), \tag{9.18}$$

if we demand that it converges at $x = 0$.

Since in general

$$ {}_1F_1(l + 1 - \lambda; 2l + 2; x) \xrightarrow[x\to\infty]{} e^x, \tag{9.19}$$

the function $u(x)$ diverges for large x unless $l + 1 - \lambda$ is a negative integer or zero, in which case the confluent hypergeometric function is a polynomial. This means that λ must be a positive integer, which we denote by N,

$$\lambda = N, \quad N = 1, 2, 3, \ldots \tag{9.20}$$

Hence, with $\sqrt{\mu} = 1/2$, Eq. (9.13) becomes

$$\mu = \frac{1}{4} = -\frac{E}{2m}\frac{N^2}{(Z\alpha)^2} \tag{9.21}$$

and we find again that the energy spectrum is discrete, with

$$E_N = -\frac{m}{2}\frac{(Z\alpha)^2}{N^2}, \quad N = 1, 2, 3, \ldots \tag{9.22}$$

The variable x is now

$$x = 2\left(\frac{Z\alpha m}{N}\right) r. \tag{9.23}$$

The spherical harmonics are normalized to unity, so the normalization of the wave function (9.9) is

$$\int |R(r)|^2 r^2 dr = \int |u(r)|^2 dr = 1, \tag{9.24}$$

where $u(r) = rR(r)$. Introduce the function of the dimensionless variable x

$$u(x) = \sqrt{\frac{N}{2Z\alpha m}}\, u(r), \tag{9.25}$$

with

$$\int |u(x)|^2 dx = 1. \tag{9.26}$$

The radial solution, which is regular at the origin and falls off exponentially at infinity, is

$$R_{Nl}(r) = \frac{u(r)}{r} = \left(\frac{2Z\alpha m}{N}\right)^{\frac{3}{2}} \frac{u_{Nl}(x)}{x} \tag{9.27}$$

where

$$u_{Nl}(x) = A_{Nl} x^{l+1} e^{-\frac{1}{2}x} {}_1F_1(-N+l+1; 2l+2; x), \tag{9.28}$$

and A_{Nl} is the normalization constant. The normalization constant is calculated in Appendix B:

$$A_{Nl} = \frac{1}{(2l+1)!} \sqrt{\frac{(N+l)!}{(N-l-1)!(2N)}}. \tag{9.29}$$

To obtain a polynomial solution we must have $N \geq l + 1$.

9.2 Laguerre Polynomials and Associated Laguerre Functions

We note that Eq. (9.16) has the form

$$xu''(x) + (p+1-x)u'(x) + (q-p)u(x) = 0, \tag{9.30}$$

where p and q are integers. Since we are looking for polynomial solutions, we can define a new function $v(x)$ by

$$u(x) = \frac{d^p v(x)}{dx^p}. \tag{9.31}$$

Clearly, if $u(x)$ is a polynomial, then $v(x)$ is also a polynomial.

On substituting this expression into Eq. (9.30) we find, after some rearrangement,

$$\frac{d^p}{dx^p}[xv''(x) + (1-x)v'(x) + qv(x)] = 0. \tag{9.32}$$

We see that if $v(x)$ satisfies *Laguerre's equation* (also a confluent hypergeometric equation)

$$xv''(x) + (1-x)v'(x) + qv(x) = 0, \tag{9.33}$$

then $u(x)$ is given by Eq. (9.31). The solution that is regular at the origin is

$$v(x) = A_q {}_1F_1(-q, 1; x). \qquad (9.34)$$

We choose $A_q = q!$ in Eq. (9.34) and obtain for the polynomial solution of Laguerre's equation

$$L_q(x) = q! {}_1F_1(-q, 1; x) = e^x \frac{d^q}{dx^q}(x^q e^{-x}), \qquad (9.35)$$

see Appendix A. This function is the *Laguerre polynomial of order q*.

From Eq. (9.31) we see that the polynomial solution to Eq. (9.30) is obtained by differentiating $L_q(x)$ p times. We denote these solutions by

$$L_q^p(x) = \frac{d^p L_q(x)}{dx^p}. \qquad (9.36)$$

These functions are known as the *associated Laguerre polynomials*. It is a polynomial of degree $q - p$. By carrying out the differentiation explicitly, we find

$$L_q^p(x) = \frac{q!(-q)_p}{p!} \sum_{k=0}^{\infty} \frac{(-q+p)_k}{k!(p+1)_k} x^k, \qquad (9.37)$$

where the *Pochhammer symbols* $(a)_n$ are defined in Appendix A.

This series is seen to describe the confluent hypergeometric function. We generalize this result to include noninteger q and p and define the *associated Laguerre function* by

$$L_q^p(x) = (-1)^p \frac{[\Gamma(q+1)]^2}{\Gamma(p+1)\Gamma(q-p+1)} {}_1F_1(-q+p, p+1; x). \qquad (9.38)$$

From Eq. (9.28) we see that the radial functions for bound states in a Coulomb potential are

$$R_{Nl}(r) = N_{Nl} \left(\frac{2Z\alpha m}{N}\right)^{\frac{3}{2}} x^l e^{-\frac{1}{2}x} L_{N+1}^{2l+1}(x), \qquad (9.39)$$

where N_{Nl} is a normalization constant. In Appendix B we calculate

$$N_{Nl} = -\sqrt{\frac{(N-l-1)!}{(N+l)!^3(2N)}}. \qquad (9.40)$$

Notes on Chapter 9

The analysis of the Schrödinger equation is conventional, e.g., Mertzbacher [Mer61]. The hypergeometric functions and the special functions which arise in different physical contexts are reviewed, for example, in the book by Seaborn [Sea80].

Chapter 10
Non-Relativistic Hydrogenic Atoms with Spin

In 1922 Stern and Gerlach passed a beam of silver atoms through an inhomogeneous magnetic field, and found that it was split into two components. Goudsmit and Uhlenbeck proposed in 1925, on the basis of atomic spectra, that the electron was characterized by an additional quantum number with value $1/2$, which would also explain the results of the Stern–Gerlach experiment. Pauli [Pau26] developed a two-component formalism, which took the spin of the electron into account. The two-component states transform as *spinors*, see [BJ53] or [FK05].

Our treatment of spin below has the benefit of hindsight. We present in this chapter a complete theory of the non-relativistic hydrogenic atom with a spinning electron. We thereby correct the long-held assumption that this theory is somehow incomplete, needing the insight provided by the Dirac theory to understand it. This is because the spin is a completely non-relativistic phenomena, as has been emphasized by Lévy-Leblond [LL74]. The Dirac theory is presented in this book as an *extension* of the successful two-component Pauli formalism.

We use the operator K_{nr} to characterize the states, which was introduced by Biedenharn and Louck [BL81] in 1981 as the non-relativistic form of an operator that was used by Dirac [Dir53]. We then show how we are led to a factorized form of the Pauli Hamiltonian, which allows a solution of the problem analogously to the harmonic oscillator. This method, which was developed for the treatment of a great variety of problems by Infeld [Inf41], and Hull and Infeld [HI51] in 1951, then suggests a supersymmetric interpretation, which is developed in the next chapter.

10.1 Spin Variables, the Pauli Hamiltonian
and Factorization

We consider a spin-$\frac{1}{2}$ Pauli particle moving in a Coulomb potential. The spin is dynamically independent, and does not affect the energy levels, which are given as before by the Bohr formula. The degeneracy, however, is doubled to the correct value $2N^2$, instead of N^2. The degeneracy becomes observable when we switch on an electric or magnetic field, which leads to a splitting of the spectral lines.

We describe a free spin-$\frac{1}{2}$ particle with the help of a two-component formalism. If the spin is independent of the space-time variables, the Hamiltonian is

$$H_P \psi = E\psi, \tag{10.1}$$

where

$$\psi = \begin{pmatrix} \psi_+ \\ \psi_- \end{pmatrix}. \tag{10.2}$$

ψ_+ is the spin-up component, ψ_- the spin-down component. If the degeneracy of the spin-up levels and the degeneracy of the spin-down levels is N^2, the degeneracy of ψ is $2N^2$. The two-component Hamiltonian H_P is

$$H_P = \frac{\mathbf{p}^2}{2m} I_2, \tag{10.3}$$

where I_2 is the unit 2×2 matrix, and H_P is called the Pauli Hamiltonian. This just says that spin-up particles and spin-down particles have exactly the same dynamics.

If p is considered as a Hermitian operator in the two-dimensional complex space then it may be expanded in terms of a basis of this space,

$$\boldsymbol{p} = p_i \sigma_i. \tag{10.4}$$

If we are to have the usual relation between the square of this vector and the scalar vector norm we must have

$$\begin{aligned} (\boldsymbol{\sigma} \cdot \boldsymbol{p})(\boldsymbol{\sigma} \cdot \boldsymbol{p}) &= (\sigma_1)^2 p_1^2 + (\sigma_2)^2 p_2^2 + (\sigma_3)^2 p_3^3 \\ &\quad + \{\sigma_1, \sigma_2\} p_1 p_2 + \{\sigma_1, \sigma_3\} p_1 p_3 + \{\sigma_2, \sigma_3\} p_2 p_3 \\ &= p_1^2 + p_2^2 + p_3^2 = \boldsymbol{p}^2. \end{aligned} \tag{10.5}$$

This holds if

$$\{\sigma_i, \sigma_j\} = \sigma_i \sigma_j + \sigma_j \sigma_i = 2\delta_{ij}, \tag{10.6}$$

where juxtaposition signifies matrix multiplication. Matrices which satisfy such a relation are said to form a *Clifford algebra*. Since space is three-dimensional, there are exactly three such matrices.

Let

$$\sigma_1 \sigma_2 \sigma_3 = X. \tag{10.7}$$

Then

$$X^2 = \sigma_1 \sigma_2 \sigma_3 \sigma_1 \sigma_2 \sigma_3 = \sigma_2 \sigma_3 \sigma_2 \sigma_3 = -I_2, \tag{10.8}$$

and

$$X = \sigma_1 \sigma_2 \sigma_3 = iI_2. \tag{10.9}$$

From this we obtain

$$\sigma_1\sigma_2 = i\sigma_3, \tag{10.10}$$

and in general

$$\sigma_i\sigma_j = i\epsilon_{ijk}\sigma_k. \tag{10.11}$$

Then

$$[\sigma_i, \sigma_j] = \sigma_i\sigma_j - \sigma_j\sigma_i = 2i\epsilon_{ijk}\sigma_k. \tag{10.12}$$

In the approach to theoretical physics called *geometrical algebra* the Clifford algebra structure plays a central role; see references in the notes to this chapter.

The commonly used representation for the Pauli matrices σ_i is

$$\sigma_1 = \begin{pmatrix} 0 & 1 \\ 1 & 0 \end{pmatrix}, \quad \sigma_2 = \begin{pmatrix} 0 & -i \\ i & 0 \end{pmatrix}, \quad \sigma_3 = \begin{pmatrix} 1 & 0 \\ 0 & -1 \end{pmatrix}. \tag{10.13}$$

The Pauli matrices obviously satisfy the identity

$$\sigma_i\sigma_j = \frac{1}{2}\{\sigma_i, \sigma_j\} + \frac{1}{2}[\sigma_i, \sigma_j] = \delta_{ij} + i\epsilon_{ijk}\sigma_k. \tag{10.14}$$

This may also be written as

$$(\boldsymbol{\sigma}\cdot\mathbf{A})(\boldsymbol{\sigma}\cdot\mathbf{B}) = \mathbf{A}\cdot\mathbf{B} + i\boldsymbol{\sigma}\cdot(\mathbf{A} \times \mathbf{B}), \tag{10.15}$$

where $\mathbf{A}$ and $\mathbf{B}$ are any two vectors which commute with $\boldsymbol{\sigma}$. With the help of this identity the Hamiltonian can be written as

$$H_P = \frac{(\boldsymbol{\sigma}\cdot\mathbf{p})(\boldsymbol{\sigma}\cdot\mathbf{p})}{2m}. \tag{10.16}$$

The advantage of writing the Hamiltonian as in Eq. (10.16) can be seen when we consider an electron in a magnetic field. The presence of the magnetic field can be accounted for by the rule of *minimal substitution*:

$$\mathbf{p} \to \boldsymbol{\pi} = \mathbf{p} - e\mathbf{A}, \tag{10.17}$$

where $\mathbf{A}$ is the vector potential (in this paragraph *only*), which satisfies $\nabla \times \mathbf{A} = \mathbf{B}$, and $\mathbf{B}$ is the magnetic field. The Hamiltonian becomes

$$H_P \to H_{\text{mag}} = \frac{1}{2m}(\boldsymbol{\sigma}\cdot\boldsymbol{\pi})(\boldsymbol{\sigma}\cdot\boldsymbol{\pi}) = \frac{\boldsymbol{\pi}^2}{2m} + \frac{i}{2m}\boldsymbol{\sigma}\cdot\boldsymbol{\pi} \times \boldsymbol{\pi}. \tag{10.18}$$

We now have

$$[\pi_i, \pi_j] = [p_i - eA_i, p_j - eA_j] = ie(\nabla_i A_j - \nabla_j A_i) = (ie)\epsilon_{ijk}B_k, \tag{10.19}$$

and

$$\boldsymbol{\sigma}\cdot\boldsymbol{\pi}\times\boldsymbol{\pi} = \frac{1}{2}\epsilon_{mij}\sigma_m[\pi_i,\pi_j] = \frac{1}{2}\epsilon_{mij}\epsilon_{ijk}\sigma_m(ie)B_k = (ie)\sigma_k B_k. \quad (10.20)$$

Therefore

$$H_{\mathrm{mag}} = \frac{\pi^2}{2m} - \left(\frac{e}{m}\right)\frac{\boldsymbol{\sigma}}{2}\cdot\boldsymbol{B} = \frac{\pi^2}{2m} - g\left(\frac{e}{m}\right)\boldsymbol{S}\cdot\boldsymbol{B} = \frac{\pi^2}{2m} - \boldsymbol{\mu}\cdot\boldsymbol{B}, \quad (10.21)$$

where S_i denote the *spin* matrices, $S_i = \frac{1}{2}\sigma_i$, and the magnetic moment is μ. $\mu_B = e/(2m)$ is called the *Bohr magneton*, and g is the gyromagnetic ratio, here $g = 2$, with

$$\boldsymbol{\mu} = g\mu_B\boldsymbol{S}. \quad (10.22)$$

The spin matrices satisfy

$$[S_i, S_j] = i\epsilon_{ijk}S_k, \quad (10.23)$$

so that the S_i are the generators of an su(2)-algebra, and $\boldsymbol{S}^2 = s(s+1)$, where $s = 1/2$ is the spin of the electron.

This is in agreement with the experimental results for the interaction of the spin of the electron with a magnetic field (up to corrections due to quantum electrodynamics). The "classical" value of g is $g = 1$, where

$$\boldsymbol{\mu} = g\left(\frac{e}{2m}\right)\boldsymbol{L}, \quad (10.24)$$

and $\boldsymbol{L}$ is the orbital angular momentum. This is nowadays recognized as the value appropriate for spin-one photons. In fact, it is conjectured that $g = 1/s$ in general. The conjecture has been proved for $s \leq 2$, and for any half-integral s ([LL74] and references therein).

This method of introducing a two-component description for particles with spin one-half, and the fact that this automatically leads to a value $g = 2$, was apparently first noticed by Feynman [Sak67]. It copies Dirac's method for the relativistic case, but works just as well for non-relativistic fermions, and could have been introduced in this case. It brings out the fact that spin is inherently a non-relativistic phenomena [LL74].

The Clifford algebra structure of the Pauli matrices allows us to factorize quadratic operators. Consider the quadratic operator $\boldsymbol{L}^2 = L_1^2 + L_2^2 + L_3^2$. We can factorize this with the help of the spin operator $\boldsymbol{\sigma}$. One finds

$$\boldsymbol{L}^2 = \boldsymbol{\sigma}\cdot\boldsymbol{L}(\boldsymbol{\sigma}\cdot\boldsymbol{L} + 1), \quad (10.25)$$

which is easily verified from Eq. (10.15) and the equation $\mathbf{L} \times \mathbf{L} = i\mathbf{L}$. The total angular momentum is $\mathbf{J} = \mathbf{L} + \frac{1}{2}\boldsymbol{\sigma}$. It factorizes as

$$J^2 = \left(\boldsymbol{\sigma}\cdot\mathbf{L} + \frac{3}{2}\right)\left(\boldsymbol{\sigma}\cdot\mathbf{L} + \frac{1}{2}\right). \tag{10.26}$$

We now introduce, following Biedenharn and Louck [BL81], the non-relativistic analog of Dirac's operator [Dir53]:

$$K_{nr} = -(\boldsymbol{\sigma}\cdot\mathbf{L} + 1) = \mathbf{L}^2 - \mathbf{J}^2 - 1/4. \tag{10.27}$$

We can then express these factorizations as

$$\mathbf{L}^2 = K_{nr}(K_{nr} + 1), \tag{10.28}$$

and

$$\mathbf{J}^2 = \left(K_{nr} + \frac{1}{2}\right)\left(K_{nr} - \frac{1}{2}\right) = K_{nr}^2 - 1/4. \tag{10.29}$$

Since K_{nr} commutes with $\mathbf{J}^2$ and $\mathbf{L}^2$ this yields, when applied to a simultaneous eigenvector, $K_{nr} \to \kappa$, $\mathbf{L}^2 \to l(\kappa)(l(\kappa) + 1)$, $\mathbf{J}^2 \to j(\kappa)(j(\kappa) + 1)$ with

$$l(\kappa) = \begin{cases} \kappa, & \kappa \text{ positive} \\ |\kappa| - 1, & \kappa \text{ negative} \end{cases} \tag{10.30}$$

and

$$j(\kappa) = |\kappa| - 1/2. \tag{10.31}$$

Since $j = 1/2, 3/2, \ldots$ we find

$$\kappa = \pm1, \pm2, \ldots (0 \text{ excluded}). \tag{10.32}$$

10.2 A Theorem Concerning the Anticommutation of K

The operator K_{nr} possesses the important property of *anticommuting* with the operator $\boldsymbol{\sigma}\cdot\boldsymbol{v}$, where $\boldsymbol{v}$ is any vector that is perpendicular to $\mathbf{L}$. In fact, a theorem of Biedenharn and Louck [Bie83] states:

Theorem. Suppose $\mathbf{v}$ is a vector with respect to the angular momentum $\mathbf{L}$, i.e.,

$$[L_i, v_j] = i\epsilon_{ijk}v_k, \tag{10.33}$$

or, in vector notation,

$$\mathbf{v} \times \mathbf{L} + \mathbf{L} \times \mathbf{v} = 2i\mathbf{v}. \tag{10.34}$$

Suppose also that this vector is perpendicular to $\mathbf{L}$,

$$\mathbf{L}\cdot\mathbf{v} = \mathbf{v}\cdot\mathbf{L} = 0. \tag{10.35}$$

Then K_{nr} anticommutes with the operator $(\boldsymbol{\sigma}\cdot\mathbf{v})$, which is a scalar with respect to the total angular momentum $\mathbf{J}$, i.e., it commutes with $\mathbf{J} = \mathbf{L} + \frac{1}{2}\boldsymbol{\sigma}$.

Proof.

$$\begin{aligned}
\{K_{nr}, (\boldsymbol{\sigma}\cdot\mathbf{v})\} &= K_{nr}(\boldsymbol{\sigma}\cdot\mathbf{v}) + (\boldsymbol{\sigma}\cdot\mathbf{v})K_{nr} \\
&= -(\boldsymbol{\sigma}\cdot\mathbf{L} + 1)(\boldsymbol{\sigma}\cdot\mathbf{v}) - (\boldsymbol{\sigma}\cdot\mathbf{v})(\boldsymbol{\sigma}\cdot\mathbf{L} + 1) \\
&= -(\boldsymbol{\sigma}\cdot\mathbf{L})(\boldsymbol{\sigma}\cdot\mathbf{v}) - (\boldsymbol{\sigma}\cdot\mathbf{v})(\boldsymbol{\sigma}\cdot\mathbf{L}) - 2(\boldsymbol{\sigma}\cdot\mathbf{v}) \\
&= -i\boldsymbol{\sigma}\cdot(\mathbf{L}\times\mathbf{v} + \mathbf{v}\times\mathbf{L}) - 2(\boldsymbol{\sigma}\cdot\mathbf{v}) = 0. \tag{10.36}
\end{aligned}$$

In the framework of the conditions of the theorem we have also

$$K_{nr}(\boldsymbol{\sigma}\cdot\mathbf{v}) = \frac{i}{2}\,\boldsymbol{\sigma}\cdot(\mathbf{v}\times\mathbf{L} - \mathbf{L}\times\mathbf{v}). \tag{10.37}$$

Important special cases result when we take for the vector $\mathbf{v}$ either $\hat{r}$ (unit radial vector), $\mathbf{p}$ (linear momentum) or $\mathcal{A}$ (Laplace vector).

10.3 Pauli Spinors

A complete set of mutually commuting operators is $\{H, \mathbf{L}^2, L_3, \sigma_3\}$. This set has energy eigenstates of the form

$$|\psi> = |Nlm> \otimes |\tfrac{1}{2}\mu> . \tag{10.38}$$

An alternative complete set is $\{H, \mathbf{J}^2, J_3, \frac{1}{2}\sigma_3\}$, with $\mathbf{J} = \mathbf{L} + \frac{1}{2}\boldsymbol{\sigma}$, and $J_3 = L_3 + \frac{1}{2}\mu$. In the position representation the states are

$$\psi(r, \theta, \phi) = R_{Nl}(r)\mathcal{Y}^{(l\frac{1}{2})jm}(\theta, \varphi), \tag{10.39}$$

because the Hamiltonian is rotationally invariant, where $\mathcal{Y}^{(l\frac{1}{2})jm}(\theta, \varphi)$ are the Pauli central field spinors: With the notation

$$\xi^1 = \left|\tfrac{1}{2}, \tfrac{1}{2}>\right. = \begin{pmatrix} 1 \\ 0 \end{pmatrix}, \quad \xi^{-1} = \left|\tfrac{1}{2}, -\tfrac{1}{2}>\right. = \begin{pmatrix} 0 \\ 1 \end{pmatrix}, \tag{10.40}$$

the two-component spinor $\mathcal{Y}^{(l\frac{1}{2})jm}(\theta, \varphi)$ is given by combining the angular momenta:

$$\mathcal{Y}^{(l\frac{1}{2})jm}(\theta, \phi) = \sum_{\mu} C\left(l\tfrac{1}{2}j; m - \mu, \mu\right) Y^l_{\mu-m}(\theta, \phi)\xi^\mu. \tag{10.41}$$

Here the $Y_m^l(\theta, \phi)$ are the spherical harmonics

$$Y_m^l(\theta, \varphi) = \sqrt{\frac{2l+1}{4\pi} \frac{(l-m)!}{(l+m)!}} \frac{(-e^{i\phi} \sin \theta)^m}{2^l l!} \left(\frac{d}{d \cos \theta}\right)^{l+m} (\cos^2 \theta - 1)^l.$$

(10.42)

The coefficients $C(l\frac{1}{2}j; m - \mu, \mu)$ are the Clebsch–Gordan coefficients.

The choice of the sign in the following results is a consequence of the phase conventions we make for the Clebsch–Gordan coefficients. The Clebsch–Gordan table with the Condon–Shortley phase conventions is

j/m	$m = \frac{1}{2}$	$m = -\frac{1}{2}$
$j = l + \frac{1}{2}$	$\sqrt{\frac{l+m+\frac{1}{2}}{2l+1}}$	$\sqrt{\frac{l-m+\frac{1}{2}}{2l+1}}$
$j = l - \frac{1}{2}$	$-\sqrt{\frac{l-m+\frac{1}{2}}{2l+1}}$	$\sqrt{\frac{l+m+\frac{1}{2}}{2l+1}}$

We now take as a complete set of mutually commuting operators

$$\{H, K_{nr}, \mathbf{J}^2, J_3\}.$$

(10.43)

The Pauli central field spinors may now be fully characterized by a notation that employs only κ and m:

$$\chi_m^\kappa = \mathcal{Y}^{[l(\kappa)\frac{1}{2})]j(\kappa)m},$$

(10.44)

where for each prescribed integer κ from the set (10.32) the quantum number m takes on the values

$$m = |\kappa| - \frac{1}{2}, |\kappa| - \frac{1}{3}, \ldots, -|\kappa| + \frac{1}{2}.$$

(10.45)

10.4 Concerning the Operator $(\boldsymbol{\sigma} \cdot \hat{\mathbf{r}})$

Consider the effect of the operator $(\boldsymbol{\sigma} \cdot \hat{\mathbf{r}})$ on $R_{Nl}(r)\chi_m^\kappa$. This operator has no effect on $R_{Nl}(r)$, it suffices to consider $(\boldsymbol{\sigma} \cdot \hat{\mathbf{r}})\chi_m^\kappa$. We want to establish

$$(\boldsymbol{\sigma} \cdot \hat{\mathbf{r}})\chi_m^\kappa = -\chi_m^{-\kappa}.$$

(10.46)

We note that $(\boldsymbol{\sigma} \cdot \hat{\mathbf{r}})$ is a scalar operator with respect to rotations, so that $(\boldsymbol{\sigma} \cdot \hat{\mathbf{r}})\chi_m^\kappa$ has the same j and m as χ_m^κ. But since $(\boldsymbol{\sigma} \cdot \hat{\mathbf{r}})$ anticommutes with K_{nr} (see theorem of Section 10.2), the result of $(\boldsymbol{\sigma} \cdot \hat{\mathbf{r}})\chi_m^\kappa$ must be proportional to $\chi_m^{-\kappa}$. That is,

$$(\boldsymbol{\sigma} \cdot \hat{\mathbf{r}})\chi_m^\kappa = c\chi_m^{-\kappa}.$$

(10.47)

Since $(\boldsymbol{\sigma}\cdot\hat{\mathbf{r}})$ is an Hermitian operator, c is a real number. Using (10.15),

$$(\boldsymbol{\sigma}\cdot\hat{\mathbf{r}})(\boldsymbol{\sigma}\cdot\hat{\mathbf{r}}) = 1, \tag{10.48}$$

we have $|c|^2 = 1$ and $c = \pm 1$. The sign is fixed by choosing $\hat{\mathbf{r}} = \hat{\boldsymbol{e}}_3 = (0,0,1)$, which is equivalent to taking $\theta = 0$. From Eq. (10.42)

$$Y_m^l(\hat{\boldsymbol{e}}_3) = \sqrt{\frac{2l+1}{4\pi}}\,\delta_{m0}, \tag{10.49}$$

so we have

$$\chi_m^\kappa(\hat{\mathbf{r}}) = \sqrt{\frac{2l(\kappa)+1}{4\pi}}\,C(l(\kappa)\tfrac{1}{2}j(\kappa);0m)\,\xi^m. \tag{10.50}$$

Equation (10.46) becomes

$$2m(\sqrt{2l_A+1})\,C\left(l_A\tfrac{1}{2}j_A;0m\right) = c\sqrt{2l_B+1}\,C\left(l_B\tfrac{1}{2}j_B;0m\right), \tag{10.51}$$

where $l_A = l(\kappa)$, $j_A = j(\kappa)$ and $l_B = l(-\kappa)$, $j_B = j(-\kappa)$. From this it follows that

$$c = 2m\sqrt{\frac{2l_A+1}{2l_B+1}}\,\frac{C\left(l_A\tfrac{1}{2}j_A;0m\right)}{C\left(l_B\tfrac{1}{2}j_B;0m\right)}. \tag{10.52}$$

When $m = \tfrac{1}{2}$ we have $2m = 1$, and

$$c = \sqrt{\frac{2l_A+1}{2l_B+1}}\,\frac{C(l_A\tfrac{1}{2}j_A;0m)}{C(l_B\tfrac{1}{2}j_B;0m)} = -1. \tag{10.53}$$

When $m = -\tfrac{1}{2}$ we have $2m = -1$, and

$$c = -\sqrt{\frac{2l_A+1}{2l_B+1}}\,\frac{C(l_A\tfrac{1}{2}j_A;0m)}{C(l_B\tfrac{1}{2}j_B;0m)} = -1. \tag{10.54}$$

In both cases we have $c = -1$, which proves (10.46). From (10.46) we get, by applying $(\boldsymbol{\sigma}\cdot\hat{\mathbf{r}})$, also

$$(\boldsymbol{\sigma}\cdot\hat{\mathbf{r}})\chi_m^{-\kappa} = -\chi_m^\kappa. \tag{10.55}$$

10.5 The Key Equation: Concerning the Operator $(\boldsymbol{\sigma}\cdot\mathcal{A})$

Applying similar considerations to the operator $A_{nr} = (\boldsymbol{\sigma}\cdot\mathcal{A})$ leads to a key equation of the whole book. We get an equation relating solutions of positive κ to solutions of negative κ, which is the essence of the application of supersymmetry, both in the relativistic and the non-relativistic cases. This equation also involves an expression for the *eccentricity*, which is a key concept of the book, again for the relativistic as well as the non-relativistic cases.

Using the definition

$$\psi_{E\kappa m} = R_{E,l(\kappa)}\chi_m^\kappa, \tag{10.56}$$

and the fact that $(\boldsymbol{\sigma}\cdot\mathbf{A})$ anticommutes with K_{nr}, one finds the result that

$$(\boldsymbol{\sigma}\cdot\mathbf{A})\psi_{E\kappa m} = -\alpha(E,\kappa)\psi_{E,-\kappa,m}, \tag{10.57}$$

where $\alpha(E,\kappa)$ is a constant.

Now remember Eq. (8.42):

$$|\mathbf{A}|^2 + \mathbf{L}^2 + 1 = \mathcal{N}^2. \tag{10.58}$$

We use also the factorization of $\mathbf{L}^2$: $\mathbf{L}^2 = K_{nr}(K_{nr}+1)$, and

$$(\boldsymbol{\sigma}\cdot\boldsymbol{A}) = |\mathbf{A}|^2 + i\boldsymbol{\sigma}\cdot\mathbf{A}\times\mathbf{A} = |\mathbf{A}|^2 - \boldsymbol{\sigma}\cdot\mathbf{L} = |\mathbf{A}|^2 + 1 + K_{nr}. \tag{10.59}$$

Here we have used

$$\boldsymbol{\sigma}\cdot(\mathbf{A}\times\mathbf{A}) = \frac{1}{2}\epsilon_{ijk}\sigma_k[A_j, A_k] = \frac{i}{2}\epsilon_{ijk}\epsilon_{jkm}\sigma_k L_m = i\boldsymbol{\sigma}\cdot\mathbf{L}. \tag{10.60}$$

Inserting this to Eq. (10.58) gives us

$$(\boldsymbol{\sigma}\cdot\mathbf{A})^2 = \mathcal{N}^2 - K_{nr}^2. \tag{10.61}$$

Hence

$$(\boldsymbol{\sigma}\cdot\mathbf{A})^2\psi_{E\kappa m} = \alpha(E,\kappa)\alpha(E,-\kappa)\psi_{E,\kappa m} = (\mathcal{N}^2 - \kappa^2)\psi_{E\kappa m}, \tag{10.62}$$

and

$$\alpha(E,\kappa) = \sqrt{N^2 - \kappa^2}. \tag{10.63}$$

For $\mathbf{A} = (\mathcal{N}^2)^{\frac{1}{2}}\boldsymbol{A}$ we get

$$(\boldsymbol{\sigma}\cdot\boldsymbol{A})\psi_{N\kappa m} = -\sqrt{1 - \kappa^2/N^2}\,\psi_{N,-\kappa,m}. \tag{10.64}$$

We shall see, in Section 10.7, that this is the expression for the eccentricity.

10.6 The Factorization Method

In this section we find a factorization property of the operator A_{nr}, which will turn out to be equivalent to the results of the supersymmetric analysis in the next chapter.

We use Eq. (10.37) to rewrite $A_{nr} = (\boldsymbol{\sigma} \cdot \mathcal{A})$ as

$$(\boldsymbol{\sigma} \cdot \mathcal{A}) = \frac{-i}{Z\alpha m} \frac{i}{2} \boldsymbol{\sigma} \cdot (\mathbf{p} \times \mathbf{L} - \mathbf{L} \times \mathbf{p}) - \boldsymbol{\sigma} \cdot \hat{\mathbf{r}} = \frac{i}{Z\alpha m}(\boldsymbol{\sigma} \cdot \mathbf{p})K_{nr} - \boldsymbol{\sigma} \cdot \hat{\mathbf{r}}. \tag{10.65}$$

We now express A_{nr} in spherical polar coordinates:

$$\boldsymbol{\sigma} \cdot \mathbf{p} = (\boldsymbol{\sigma} \cdot \hat{\mathbf{r}})(\boldsymbol{\sigma} \cdot \hat{\mathbf{r}})\boldsymbol{\sigma} \cdot \mathbf{p} = \boldsymbol{\sigma} \cdot \hat{\mathbf{r}} [\hat{r} \cdot \mathbf{p} + i\boldsymbol{\sigma} \cdot (\hat{r} \times \mathbf{p})]$$

$$= \boldsymbol{\sigma} \cdot \hat{\mathbf{r}} [\hat{r} \cdot \mathbf{p} + ir^{-1}\boldsymbol{\sigma} \cdot \mathbf{L}] = \boldsymbol{\sigma} \cdot \hat{\mathbf{r}} [\hat{r} \cdot \mathbf{p} - ir^{-1}(K_{nr} + 1)]$$

$$= -i\boldsymbol{\sigma} \cdot \hat{\mathbf{r}} \left[\frac{\partial}{\partial r} + \frac{K_{nr} + 1}{r} \right] = -i\boldsymbol{\sigma} \cdot \hat{\mathbf{r}} \left[ip_r + \frac{K_{nr}}{r} \right], \tag{10.66}$$

where, according to Eq. (9.6),

$$p_r = -\frac{i}{r} \frac{\partial}{\partial r} r. \tag{10.67}$$

Inserting this into Eq. (10.65) we obtain:

$$A_{nr} = (\boldsymbol{\sigma} \cdot \mathcal{A}) = (\boldsymbol{\sigma} \cdot \hat{\mathbf{r}}) \left[\frac{K_{nr}}{Z\alpha m} \left(ip_r + \frac{K_{nr}}{r} \right) - 1 \right]. \tag{10.68}$$

This result, when combined with Eqs. (10.64) and remembering the sign change due to Eq. (10.46), leads directly to the radial differential equation in the form

$$\left[\frac{\kappa}{Z\alpha m} \left(ip_r + \frac{\kappa}{r} \right) - 1 \right] R_{N,l(\kappa)}(r) = \sqrt{1 - \frac{\kappa^2}{N^2}} \ R_{N,l(-\kappa)}(r). \tag{10.69}$$

When κ is positive, this defines a lowering operator on l: $l(\kappa) = \kappa$, $l(-\kappa) = \kappa - 1 = l - 1$,

$$\left[\frac{l}{Z\alpha m} \left(\frac{d}{dr} + \frac{l+1}{r} \right) - 1 \right] R_{Nl}(r) = \sqrt{1 - \frac{l^2}{N^2}} \ R_{N,l-1}(r). \tag{10.70}$$

When κ is negative it defines a raising operator on l: $l(\kappa) = |\kappa| - 1$, $\kappa = -(l+1)$, $l(-\kappa) = |\kappa| = l + 1$,

$$\left[-\frac{(l+1)}{Z\alpha m} \left(\frac{d}{dr} - \frac{l}{r} \right) - 1 \right] R_{Nl}(r) = \sqrt{1 - \frac{(l+1)^2}{N^2}} \ R_{N,l+1}(r). \tag{10.71}$$

These equations may be written as

$$A_{l+1}^- R_{N,l+1}(r) = g_{l+1} R_{Nl}(r), \quad A_{l+1}^+ R_{Nl} = g_{l+1} R_{N,l+1}(r), \tag{10.72}$$

with

$$g_l = Z\alpha m \left(\frac{1}{l^2} - \frac{1}{N^2}\right)^{\frac{1}{2}}, \quad A_l^{\mp} = \pm i p_r + \frac{l}{r} - \frac{Z\alpha m}{l}, \tag{10.73}$$

and $A^+ = (A^-)^\dagger$.

Applying the raising operator to $R_{N,l}(r)$, and then the lowering operator to the resulting $R_{N,l+1}(r)$, yields

$$A_{l+1}^- A_{l+1}^+ R_{Nl} = g_{(l+1)}^2 R_{Nl}, \tag{10.74}$$

or

$$\left[\frac{d}{dr} + \frac{(l+2)}{r} - \frac{Z\alpha m}{l+1}\right]\left[-\frac{d}{dr} + \frac{l}{r} - \frac{Z\alpha m}{l+1}\right] R_{Nl}(r)$$

$$= (Z\alpha m)^2 \left[\frac{1}{(l+1)^2} - \frac{1}{N^2}\right] R_{Nl}(r). \tag{10.75}$$

This can be simplified to

$$\left[\frac{1}{r^2}\frac{d}{dr}r^2\frac{d}{dr} - \frac{l(l+1)}{r^2} + \frac{2Z\alpha m}{r} - 2mE_N\right] R_{Nl}(r) = 0, \tag{10.76}$$

with E_N given by Eq. (8.48). We have recovered the Schrödinger equation for the radial part of the Pauli wave function. But since it has now been factorized into the product of two factors, which raise and lower l without changing the energy, we can now explain the degeneracy in l, and the degeneracy $2N^2$ of an N-multiplet.

The tower of states terminates when $A_{l+1}^+ R_{Nl} = 0$, that is, when $g_{l+1} = 0$. From Eq. (10.73) we see that this implies $N = l+1$. From the equation $A_N^+ R_{Nl} = 0$ we further have

$$\left[-\frac{d}{dr} + \frac{N-1}{r} - \frac{1}{N}\right] R_{N,N-1}(r) = 0, \tag{10.77}$$

and hence

$$R_{N,N-1}(r) = N_{N,N-1}\left[\frac{2(Z\alpha m)r}{N}\right]^{N-1} e^{-\frac{Z\alpha m}{N}r}. \tag{10.78}$$

The radial wave function is nodeless. These orbits are called *"circular"*. For the ground state ($N = 1$) we have

$$R_{10}(r) = (\text{constant})e^{-(Z\alpha m)r}. \tag{10.79}$$

All other eigenfunctions in the N-multiplet can be generated from the explicit result for the eigenfunction with $l = N - 1$. Taking $\kappa = N - 1$ (since $l(N - 1) = N - 1$) we can use this same eigenfunction in Eq. (10.70), which now functions as a lowering operator, generating the eigenfunction for $l = N - 2$. Iterating this procedure yields all the eigenfunctions in an N-multiplet.

To summarize: The properties of the operator A_{nr} are equivalent to a definition of the radial eigenfunctions; moreover, the eigenvalue conditions deducible from $A_{nr} \to 0$ are equivalent to the eigenvalue condition (obtained from analysis) for the confluent hypergeometric function $_1F_1(a, b; z)$. The eigenvalue condition referred to above is the condition that the confluent hypergeometric function reduces to a polynomial for the parameter a being zero or a negative integer.

10.7 The Definition of Eccentricity

In the spin-$\frac{1}{2}$ case the vector $\mathbf{p}$ is replaced by the 2×2 matrix $(\boldsymbol{\sigma} \cdot \mathbf{p})$. In the same way, the appropriate expression for the Laplace vector $\boldsymbol{\mathcal{A}}$ becomes $A_{nr} = (\boldsymbol{\sigma} \cdot \boldsymbol{\mathcal{A}})$. The eccentricity is, in the classical case, the magnitude of the Laplace vector $\boldsymbol{\mathcal{A}}$. In the case of a spin-$\frac{1}{2}$ particle the vector $\boldsymbol{\mathcal{A}}$ is replaced by the quantity $(\boldsymbol{\sigma} \cdot \boldsymbol{\mathcal{A}})$. So the corresponding statement, for spin-$\frac{1}{2}$ quantum particles, is that the eccentricity is the magnitude of the quantity $(\boldsymbol{\sigma} \cdot \boldsymbol{\mathcal{A}})$.

Since

$$e \, \psi_{N\kappa m} = \sqrt{(\boldsymbol{\sigma} \cdot \boldsymbol{\mathcal{A}})^2} \, \psi_{N\kappa m} = \sqrt{1 - \kappa^2/N^2} \, \psi_{N\kappa} \qquad (10.80)$$

we have

$$e = \sqrt{1 - \kappa^2/N^2}, \qquad (10.81)$$

with $\kappa = \pm 1, \pm 2, \ldots, \pm(N - 1), -N$. This corresponds to the operator equation

$$e_{\text{op}} = [1 - K_{nr}^2/\mathcal{N}^2]^{\frac{1}{2}}. \qquad (10.82)$$

The circular orbits have eccentricity zero, or $\kappa = -N$, and $l(\kappa) = N - 1$. Then we have the statement, both classically and in quantum mechanics, that the circular orbits are those with vanishing eccentricity.

Figure 10.1 is a picture of the orbits of hydrogen according to quantum mechanics. The circular orbits for $N = 1, 2, 3$ are shown. The orbits with non-vanishing eccentricity are seen to deviate from symmetry, to a degree proportional to their eccentricity.

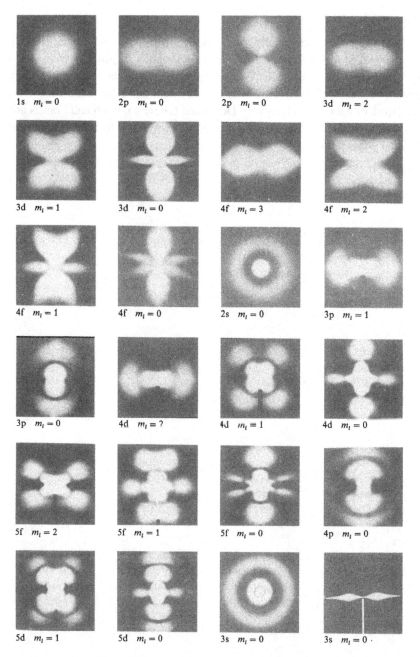

Figure 10.1. Low-lying orbits of the hydogen atom.

Notes on Chapter 10

This chapter follows essentially the work of Biedenharn and Louck [Bie62]. The importance of Clifford algebras for the description of particles with spin is especially emphasized in geometrical algebra; see Doran and Lasenby [DL03]. Spinors are discussed, at an elementary level, by Bade and Jehle [BJ53], and in the framework of geometric algebra, by Doran and Lasenby [DL03] and Francis and Kosowsky [FK05]. The factorization method was developed by Infeld [Inf41] and Hull and Infeld [HI51].

Chapter 11

Elements of Supersymmetric Quantum Mechanics

The theory of supersymmetry has its origins in quantum field theories [SDB04] and elementary particle physics [WZ74], but it is also a valuable tool for describing the solutions of many ordinary quantum mechanical systems [Wit90], [RLB90]. It is used, for example, to provide a simple proof of the positive-mass conjecture in general relativity [Wit81], and in the proof of various index theorems in differential geometry [Tak08]. The following is a summary of the most pertinent definitions that we need in our discussion of supersymmetry in non-relativistic hydrogenic atoms and, later, in the Dirac–Kepler problem.

11.1 General Considerations

Assume that we have a vector space $\mathcal{V}$ that is the direct sum of two vector spaces $\mathcal{V}_1$ and $\mathcal{V}_2$:

$$\mathcal{V} = \mathcal{V}_1 \oplus \mathcal{V}_2, \tag{11.1}$$

such that the vectors of $\mathcal{V}_1$ are even with respect to some *parity* $\mathcal{P}$, while those of $\mathcal{V}_2$ are odd. A general vector in $\mathcal{V}$ may then be written as the sum of an even and an odd vector:

$$|\nu\rangle = \begin{bmatrix} |\nu_1\rangle \\ |\nu_2\rangle \end{bmatrix} = \begin{bmatrix} |\nu_1\rangle \\ 0 \end{bmatrix} + \begin{bmatrix} 0 \\ |\nu_2\rangle \end{bmatrix}. \tag{11.2}$$

The vector space $\mathcal{V}$ is called Z_2-graded.

Similarly, a linear operator in the space $\mathcal{V}$ may be written as the sum of an even and an odd operator:

$$\Omega = \begin{bmatrix} \Omega_{11} & \Omega_{12} \\ \Omega_{21} & \Omega_{22} \end{bmatrix} = \begin{bmatrix} \Omega_{11} & 0 \\ 0 & \Omega_{22} \end{bmatrix} + \begin{bmatrix} 0 & \Omega_{12} \\ \Omega_{21} & 0 \end{bmatrix}. \tag{11.3}$$

The element Ω_{11} effects an even-even transition. Similarly, the element Ω_{22} is odd-odd. The element Ω_{12} describes the transformation of an odd state into an even one, the element Ω_{21} the transformation of an even state into an odd one. The linear operators form a graded algebra. The even and the odd operators are called homogeneous.

The set of linear operators define a superalgebra if it is closed under a generalized commutator, which, for two homogeneous operators Ω_1 and Ω_2, is the ordinary commutator between Ω_1 and Ω_2 unless both elements are odd. In the latter case, it is the anticommutator between Ω_1 and Ω_2. The definition of the generalized commutator is extended to arbitrary operators by requiring it to be a bilinear operation.

Consider the so-called s(2) superalgebra defined by the supercharges

$$Q_1 = \frac{1}{\sqrt{2}} \begin{bmatrix} 0 & A^+ \\ A^- & 0 \end{bmatrix}, \quad Q_2 = \frac{i}{\sqrt{2}} \begin{bmatrix} 0 & A^+ \\ -A^- & 0 \end{bmatrix}, \tag{11.4}$$

and the supersymmetric Hamiltonian

$$\mathcal{H} = \begin{bmatrix} A^+ A^- & 0 \\ 0 & A^- A^+ \end{bmatrix} = \begin{bmatrix} H_+ & 0 \\ 0 & H_- \end{bmatrix}, \tag{11.5}$$

where the operators A^+ and A^- are the Hermitian adjoints of each other. The operators

$$Q = \frac{1}{\sqrt{2}} (Q_1 + iQ_2) = \begin{bmatrix} 0 & 0 \\ A^- & 0 \end{bmatrix} \tag{11.6}$$

and

$$Q^\dagger = \frac{1}{\sqrt{2}} (Q_1 - iQ_2) = \begin{bmatrix} 0 & A^+ \\ 0 & 0 \end{bmatrix} \tag{11.7}$$

are the supersymmetry generators. The parity operator that is responsible for the grading is represented by the operator

$$\mathcal{P} = \begin{bmatrix} 1 & 0 \\ 0 & -1 \end{bmatrix}. \tag{11.8}$$

By explicit matrix multiplication we have

$$Q_1 Q_2 = \frac{i}{2} \begin{bmatrix} -A^+ A^- & 0 \\ 0 & A^- A^+ \end{bmatrix}, \quad Q_2 Q_1 = \frac{i}{2} \begin{bmatrix} A^+ A^- & 0 \\ 0 & -A^- A^+ \end{bmatrix}. \tag{11.9}$$

From this $\{Q_1, Q_2\} = 0$ and similarly $Q_1^2 = Q_2^2 = \frac{1}{2}\mathcal{H}$. We also have

$$\{Q, Q\} = \{Q^\dagger, Q^\dagger\} = 0, \quad \text{and} \quad \{Q, Q^\dagger\} = Q_1^2 + Q_2^2 = \mathcal{H}. \tag{11.10}$$

Therefore

$$[Q_1, \mathcal{H}] = [Q_1, Q_1^2 + Q_2^2] = [Q_1, Q_2^2] = -Q_2\{Q_1, Q_2\} + \{Q_1, Q_2\}Q_2 = 0. \tag{11.11}$$

Similarly $[Q_2, \mathcal{H}] = 0$.

Thus, the supersymmetric Hamiltonian is the sum of the squares of the supersymmetric charges, or the anticommutator of the supersymmetry generators. These are relations that characterize supersymmetric quantum mechanics, with the proviso that H_+ and H_- be closely related to the Hamiltonians of actual physical systems. We say that a system which satisfies these relations is an s(2) superalgebra, and we speak of an S(2) supersymmetry.

Assume now that the vector $|\nu\rangle$ defined by Eq. (11.2) is the eigenvector of the supersymmetric Hamiltonian with eigenvalue $\mathcal{E}$:

$$\mathcal{H}\begin{bmatrix} |\nu_1\rangle \\ |\nu_2\rangle \end{bmatrix} = \begin{bmatrix} H_+ & 0 \\ 0 & H_- \end{bmatrix}\begin{bmatrix} |\nu_1\rangle \\ |\nu_2\rangle \end{bmatrix} = \mathcal{E}\begin{bmatrix} |\nu_1\rangle \\ |\nu_2\rangle \end{bmatrix}. \tag{11.12}$$

$|\nu_1\rangle$ is then an eigenstate of the Hamiltonian H_+ with eigenvalue $\mathcal{E}$, and $|\nu_2\rangle$ is an eigenstate of H_- with the same eigenvalue.

Assuming that $|\nu_1\rangle$ and $|\nu_2\rangle$ are normalizable vectors, we have

$$|A^-|\nu_1\rangle|^2 = \langle \nu_1|A^+A^-|\nu_1\rangle = \langle A^-\nu_1|A^-\nu_1\rangle = \mathcal{E}\langle \nu_1|\nu_1\rangle,$$
$$|A^+|\nu_2\rangle|^2 = \langle \nu_2|A^-A^+|\nu_2\rangle = \langle A^+\nu_2|A^+\nu_2\rangle = \mathcal{E}\langle \nu_2|\nu_2\rangle, \tag{11.13}$$

where we have used that A^+ and A^- are the Hermitian adjoints of each other. These relations show that $\mathcal{E}$ is positive or zero. They also show that $A^-|\nu_1\rangle$ and $A^+|\nu_2\rangle$ vanish if and only if $\mathcal{E}$ is zero.

Therefore the vectors

$$Q\begin{bmatrix} |\nu_1\rangle \\ 0 \end{bmatrix} = \begin{bmatrix} 0 \\ A^-|\nu_1\rangle \end{bmatrix}, \quad Q^\dagger\begin{bmatrix} 0 \\ |\nu_2\rangle \end{bmatrix} = \begin{bmatrix} A^+|\nu_2\rangle \\ 0 \end{bmatrix} \tag{11.14}$$

vanish if $\mathcal{E}$ is zero. Otherwise, they are also eigenvectors of $\mathcal{H}$ with eigenvalue $\mathcal{E}$, because Q and $Q^\dagger$ commute with $\mathcal{H}$. Thus $A^-|\nu_1\rangle$ is an eigenvector of H_- with energy $\mathcal{E}$ and $A^+|\nu_2\rangle$ is an eigenvector of H_+, also with energy $\mathcal{E}$.

Any eigenstate of H_+ with positive energy has, accordingly, a partner that is an eigenstate of H_- with the same energy and vice versa. The eigenstates of H_+

and H_- with positive energy are paired. They are transformed into each other by the operators A^- and A^+. Thus the levels of the super-Hamiltonian are doubly degenerate. If, however, either H_+ or H_- has an eigenstate with energy zero, then there is no partner state and, hence, no degeneracy.

We have seen that an eigenstate of $\mathcal{H}$ with energy zero is annihilated by Q and $Q^\dagger$. If, conversely, an eigenstate of $\mathcal{H}$ is annihilated by Q and $Q^\dagger$, then its energy is zero. This follows from Eq. (11.10), according to which $\{Q, Q^\dagger\} = \mathcal{H}$.

For supersymmetry to be an exact symmetry, one requires that $\mathcal{H}$ have a supersymmetrically invariant ground state, i.e., a state that is annihilated by Q and $Q^\dagger$. Such a ground state has zero energy, and either $|\nu_1\rangle$ or $|\nu_2\rangle$ is zero.

We have thus seen that if supersymmetry is an exact symmetry, then the two Hamiltonians H_+ and H_- have the same set of eigenvalues, except that one of them has a zero eigenvalue that is not shared by the other Hamiltonian.

This completes our summary of elementary supersymmetric quantum mechanics. We shall now turn to its application to non-relativistic hydrogenic atoms with spin.

11.2 Supersymmetry of Non-Relativistic Hydrogenic Atoms

One way of applying supersymmetry to non-relativistic hydrogen atoms is to consider the $Z(2)$ grading to be just spin-up or spin-down. In other words an electron is written as a two-component entity with

$$|\psi^+\rangle = \begin{bmatrix} 1 \\ 0 \end{bmatrix}, \quad |\psi^-\rangle = \begin{bmatrix} 0 \\ 1 \end{bmatrix}. \tag{11.15}$$

This corresponds to a grading with $P_3 = \sigma_3$.

Consider the Pauli Hamiltonian for an atom placed in a magnetic field along the z-axis: $\boldsymbol{B} = B\boldsymbol{e}_3$. This field is generated by a magnetic potential $\boldsymbol{A}(x, y) = (A_x, A_y, 0)$. The Pauli Hamiltonian restricted to the (x, y)-plane is

$$H_{\text{mag}} = \frac{1}{2m}(\pi_x^2 + \pi_y^2) - g\mu_B B S_3, \tag{11.16}$$

where π_x, π_y stand for the abbreviations

$$\pi_x = p_x - eA_x, \quad \pi_y = p_y - eA_y, \tag{11.17}$$

g is the gyromagnetic factor and μ_B is the Bohr magneton.

Introduce the supercharges

$$Q_1 = \frac{1}{2\sqrt{m}}(-\pi_y\sigma_1 + \pi_x\sigma_2),$$

$$Q_2 = \frac{1}{2\sqrt{m}}(\pi_x\sigma_1 + \pi_y\sigma_2). \tag{11.18}$$

We may now calculate, using the properties of σ_i, Eq. (10.6),

$$Q_1Q_2 = \frac{1}{4m}(-\pi_y\pi_x + \pi_x^2\sigma_2\sigma_1 - \pi_y^2\sigma_1\sigma_2 + \pi_x\pi_y),$$

$$Q_2Q_1 = \frac{1}{4m}(-\pi_x\pi_y - \pi_y^2\sigma_2\sigma_1 + \pi_x^2\sigma_1\sigma_2 + \pi_y\pi_x). \tag{11.19}$$

These two equations may be expressed more compactly as

$$\{Q_1, Q_2\} = \frac{1}{4m}(\pi_x^2 - \pi_y^2)\{\sigma_1, \sigma_2\} = 0. \tag{11.20}$$

We also have

$$Q_1^2 = \frac{1}{4m}(\pi_x^2 + \pi_y^2) - \frac{1}{4m}[\pi_x, \pi_y]\sigma_1\sigma_2,$$

$$Q_2^2 = \frac{1}{4m}(\pi_x^2 + \pi_y^2) + \frac{1}{4m}[\pi_x, \pi_y]\sigma_2\sigma_1. \tag{11.21}$$

Since $\sigma_1\sigma_2 = -\sigma_2\sigma_1$ we have $Q_1^2 = Q_2^2$. For the commutators we have, according to Eq. (10.19):

$$[\pi_x, \pi_y] = ieB. \tag{11.22}$$

Inserting this in the foregoing equations, then the comparison with the formula for H, Eq. (11.16) — and the demand that $g = 2$ — leads to the equation $Q_1^2 = Q_2^2 = \frac{1}{2}H$. It follows that H commutes both with Q_1 and Q_2. Together with Eq. (11.20) we can summarize this as

$$\{Q_i, Q_j\} = H\delta_{ij}, \quad \text{and} \quad [H, Q_i] = 0. \tag{11.23}$$

In terms of $Q, Q^\dagger$ we have

$$\{Q, Q\} = \{Q^\dagger, Q^\dagger\} = [H, Q] = [H, Q^\dagger] = 0, \quad \{Q, Q^\dagger\} = H. \tag{11.24}$$

Thus we have the supersymmetry algebra s(2). We may say that the requirement $g = 2$ is equivalent to demanding that the Pauli Hamiltonian, together with the supercharges, form a supersymmetric algebra. We have attempted to make clear that it is in any case a very reasonable requirement. It does *not* require the Dirac equation for its justification, contrary to what is often claimed in the literature.

Another application is a Z_2 grading introduced in the Hilbert space of states by classifying the states to be even or odd with respect to the parity operator $P_\kappa = K_{nr}/|\kappa|$, i.e., having eigenvalues $\pm Sign(\kappa)$, respectively. Equally, linear operators can be assigned a grade. An operator is even if it commutes with P_κ, whereas operators anticommuting with P_κ are called odd. As an example of an even operator we mention H, following from the fact that K_{nr} is built from symmetry operators of H. Of course, K_{nr} itself is even as well. According to the Theorem 10.2, for any vector $\mathbf{v}$ that is perpendicular to $\mathbf{L}$, the $\mathbf{J}$ scalar $(\boldsymbol{\sigma}\cdot\mathbf{v})$ anticommutes with K_{nr}, and is hence an odd operator. This theorem supplies us with odd operators $(\boldsymbol{\sigma}\cdot\mathbf{v})$, with $\mathbf{v}$ equal to, for example, $\mathbf{p}, \mathbf{r}$ or $\mathcal{A}$.

The supercharges communicate between the subspaces with $Sign(\kappa) = \pm 1$, and can be taken to be $(\boldsymbol{\sigma}\cdot\mathcal{A})$. Since $(\boldsymbol{\sigma}\cdot\mathcal{A})$ is Hermitian $A^+ = A^- = (\boldsymbol{\sigma}\cdot\mathcal{A})$. Then we have

$$Q = \begin{bmatrix} 0 & 0 \\ (\boldsymbol{\sigma}\cdot\mathcal{A}) & 0 \end{bmatrix}, \quad Q^\dagger = \begin{bmatrix} 0 & (\boldsymbol{\sigma}\cdot\mathcal{A}) \\ 0 & 0 \end{bmatrix}. \tag{11.25}$$

When Q acts on the vector $|\nu\rangle$ then we have $(\boldsymbol{\sigma}\cdot\mathcal{A})$ acting on the positive κ component, so that $(\boldsymbol{\sigma}\cdot\mathcal{A})$ lowers $l(\kappa)$ by one unit, whereas $Q^\dagger$ acts on the negative κ component, and raises $l(\kappa)$ by one unit.

We also have

$$\mathcal{H}\begin{bmatrix} |\nu_1\rangle \\ |\nu_2\rangle \end{bmatrix} = \{Q, Q^\dagger\}\begin{bmatrix} |\nu_1\rangle \\ |\nu_2\rangle \end{bmatrix} = (\boldsymbol{\sigma}\cdot\mathcal{A})^2 \begin{bmatrix} |\nu_1\rangle \\ \nu_2 \end{bmatrix} = \mathcal{E}\begin{bmatrix} |\nu_1\rangle \\ |\nu_2\rangle \end{bmatrix}, \tag{11.26}$$

with

$$\mathcal{E} = \left(1 - \frac{\kappa^2}{N^2}\right), \tag{11.27}$$

see Eq. (10.81).

In general an eigenstate of $(\boldsymbol{\sigma}\cdot\mathcal{A})^2$ with positive energy has a partner of the same energy, but with opposite value of κ. Thus, the levels of the super-Hamiltonian are doubly degenerate. If, however, there is an eigenstate with $\mathcal{E} = 0$ then there is no partner state and no degeneracy. Supersymmetry is exact when $\mathcal{H}$ has a supersymmetrically invariant ground state, that is, a state that is annihilated by $A_{nr} = (\boldsymbol{\sigma}\cdot\mathcal{A})$.

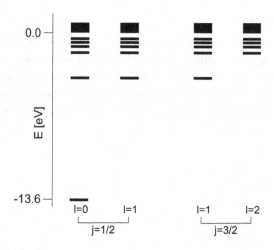

Figure 11.1.   The supersymmetry of the non-relativistic hydrogen spectrum.

The constructed S(2) supersymmetry has implications for the spectrum. The consequences of a (super-) symmetry group of the Hamiltonian is that the energy eigenspaces consist of irreducible representations of the group. The irreducible representations of S(2) are either one- or two-dimensional. States within the subspace $\mathcal{H}_\kappa$, with fixed j and m, transform irreducibly under the superalgebra s(2). The multiplets have dimensionality two, unless $\mathcal{E} = 0$, in which case they are one-dimensional.

We can now interpret the spectrum of a non-relativistic spin-$\frac{1}{2}$ particle in a Coulomb field (see Figure 11.1):

1. The degeneracies of levels with fixed j and m but with $l(\kappa) = j \pm \frac{1}{2}$ are a consequence of the s(2) supersymmetry algebra, constructed above. The level of degeneracy is two, except for states with $l = j - \frac{1}{2}$ and

$$E = -\frac{m}{2} \frac{(Z\alpha)^2}{(l+1)^2}, \qquad (11.28)$$

   which are nondegenerate.
2. The degeneracies of fixed l and m but $j = l \pm \frac{1}{2}$ are due to kinematic independence of the electron spin in the non-relativistic regime.
3. The $(2j + 1)$-fold degeneracy of states with fixed j and l is due to rotational invariance of the interaction.

Notes on Chapter 11

The relation of the factorization method to supersymmetry has been emphasized by Stahlhofen and Bleuler [SB89]. The application of supersymmetry to the non-relativistic hydrogen spectrum was performed by Tangerman and Tjon [TT93]. The various applications of supersymmetry to quantum mechanics are reviewed in Roy, Lahiri and Bagchi [RLB90] and Khare, Cooper and Sukhatma [KCS95].

Chapter 12
Sommerfeld's Derivation of the Relativistic Energy Level Formula

Sommerfeld [Som16] developed in 1916 a generalization of the Bohr model, which allowed for elliptical orbits and was consistent with the theory of relativity. In the classical relativistic extension of the problem of the motion of a particle in a Coulomb field the orbits are *precessing* ellipses. Applying the quantization conditions to these orbits leads to the *Sommerfeld formula* for the energy levels.

This was done by Sommerfeld in an attempt to explain the *fine structure* of the hydrogen atom. The fine structure is a splitting of the spectral lines into several distinct components, which is found in all atomic spectra. It can be observed only by using equipment of very high resolution since the separation, in terms of reciprocal wavelength, between adjacent components of a single spectral line is of the order of 10^{-4} times the separation between adjacent lines. This must mean that what we had thought was a single energy state of the hydrogen atom actually consists of several states that are very close together in energy.

The surprising feature of this formula is its agreement with the result of the full Dirac theory. This despite the fact that Sommerfeld used the methods of the "Old Quantum Theory", and that his treatment completely neglects the fact that the electron is a particle with spin. The agreement of these two formulas, despite their different origins, is referred to as the "Sommerfeld puzzle". In the subsequent chapters we shall look at the Dirac description of the spinning electron, and discuss the resolution of the puzzle, following the treatment of Biedenharn [Bie83].

In the present chapter we present Sommerfeld's model. It is a direct generalization of the treatment of the non-relativistic case treated in Chapter 7.

12.1 Assumptions of the Model

The relativistic Hamiltonian for an attractive Coulomb potential $-(Z\alpha)/r$, has the quadratic form

$$\left(H + \frac{Z\alpha}{r}\right)^2 = \mathbf{p}^2 + m^2. \tag{12.1}$$

This can be interpreted as the relativistic energy-momentum relation $E_{kin}^2 - \mathbf{p}^2 = m^2$.

The momentum $\mathbf{p}$ can be resolved classically into two components:

$$\mathbf{p}^2 = p_r^2 + \frac{p_\theta^2}{r^2}, \tag{12.2}$$

where the radial momentum is

$$p_r = \frac{m\dot{r}}{\sqrt{1 - v^2}}, \tag{12.3}$$

and the angular momentum is

$$p_\theta = \frac{mr^2\dot{\theta}}{\sqrt{1 - v^2}}. \tag{12.4}$$

From angular momentum conservation, the component p_θ is a constant of the motion. We define as before the variable $s = 1/r$, and note that

$$\left(\frac{p_r}{p_\theta}\right) = -\frac{ds}{d\theta}. \tag{12.5}$$

Introducing these variables into Eq. (12.1), we find that

$$\left(\frac{H + Z\alpha s}{m}\right)^2 = 1 + \left(\frac{p_\theta}{m}\right)^2 \left(\frac{ds}{d\theta}^2 + s^2\right). \tag{12.6}$$

As before, one differentiates Eq. (12.6) with respect to θ and obtains the linear differential equation

$$\frac{d^2 s}{d\theta^2} + \gamma^2(s - C) = 0, \tag{12.7}$$

where we have used the abbreviations

$$\gamma^2 = 1 - \frac{(Z\alpha)^2}{p_\theta^2} \tag{12.8}$$

and

$$C = \left(\frac{Z\alpha H}{\gamma^2 p_\theta^2} \right). \tag{12.9}$$

From Eq. (12.7) we obtain the general solution for the relativistic orbits:

$$s = \frac{1}{r} = A\cos(\gamma\theta) + B\sin(\gamma\theta) + C, \tag{12.10}$$

where A, B are the integration constants and C is given in Eq. (12.9). Defining the distance of closest approach (perihelion) to occur at $\theta = 0$ makes $B = 0$, yielding the simpler equation

$$\frac{1}{r} = A\cos(\gamma\theta) + C. \tag{12.11}$$

This result shows that the classical Kepler orbits have the form of *conic sections* — just as in the non-relativistic case — but that the correct angle variable (in terms of which the orbit is *actually* a conic section) is not θ but $\psi = \gamma\theta$. Thus, for elliptic orbits (bound states) to move from one perihelion ($\psi = 0$) to the next ($\psi = 2\pi$) requires $\theta = 2\pi/\gamma$. Since the parameter γ is less than unity, classical bound state relativistic Kepler motion takes place in an ellipse whose perihelion is advancing (that is, moving in the same sense as the orbit is transversed).

We now express the elliptic orbits in geometrical terms, using the eccentricity e. For $\psi = 0$ one has the perihelion distance $r = a(1-e)$ and, for $\psi = \pi$, the aphelion distance $r = a(1 + e)$. The orbit equation, Eq. (12.11), is then

$$\frac{1}{r} = \frac{1}{a}\left(\frac{1 + e\cos\psi}{1 - e^2} \right). \tag{12.12}$$

Except for the occurrence of $\psi = \gamma\theta$ in place of θ, one sees that the orbits are exactly of non-relativistic form. In other words, in a moving reference frame, such that the perihelion appears fixed, the orbit is non-relativistic in form.

It is now easy to derive Sommerfeld's formula by applying the Sommerfeld–Wilson quantization rules to the phase integrals for p_θ and p_r:

$$\int_{\theta=0}^{\theta=2\pi} p_\theta d\theta = 2\pi n_\theta, \tag{12.13}$$

which yields immediately $p_\theta = n_\theta$, and

$$\oint p_r dr = 2\pi n_r. \tag{12.14}$$

To evaluate the integral in Eq. (12.14) rewrite the radial momentum,

$$p_r = \frac{m\dot{r}}{\sqrt{1 - v^2}} = \frac{m}{\sqrt{1 - v^2}} \left(\frac{dr}{d\theta}\right) \dot{\theta} = \frac{p_\theta}{r^2} \left(\frac{dr}{d\theta}\right), \qquad (12.15)$$

so that we obtain

$$p_r dr = p_\theta \left(\frac{1}{r}\frac{dr}{d\theta}\right)^2 d\theta = p_\theta \frac{e^2 \gamma \sin^2 \psi}{(1 + e \cos \psi)^2} d\psi, \qquad (12.16)$$

upon using the orbit Eq. (12.12). Thus, the radial quantization condition, Eq. (12.14), takes the form

$$\left(\frac{1}{2\pi}\right) \int_{\psi=0}^{\psi=2\pi} \frac{e^2 \sin^2 \psi \, d\psi}{(1 + e \cos \psi)^2} = \frac{n_r}{\gamma n_\theta}. \qquad (12.17)$$

Notice that the left-hand side of Eq. (12.17) is identical to that for non-relativistic motion, since the integral involves only the geometry of the ellipse. It has the value

$$\frac{1}{2\pi} \int_{\psi=0}^{\psi=2\pi} \frac{e^2 \sin^2 \psi \, d\psi}{(1 + e \cos \psi)^2} = (1 - e^2)^{-\frac{1}{2}} - 1. \qquad (12.18)$$

By contrast, the right-hand side of Eq. (12.17) differs from the non-relativistic case solely by the appearance of the factor γ. We may phrase this more suggestively by saying that the sole effect of relativity in the radial quantization condition consists in replacing the orbital angular momentum, n_θ, by the "irrational angular momentum" parameter, γn_θ.

12.2 The Energies of the Bound States

It is now straightforward to obtain the formula for the energy levels. The two parameters of the orbit equation (a and e in Eq. (12.12)) are determined from the quantization conditions:

1. From Eq. (12.18) we know the eccentricity, e:

$$\frac{1}{1 - e^2} = \left(1 + \frac{n_r}{\gamma n_\theta}\right)^2. \qquad (12.19)$$

2. From Eq. (12.12) we have

$$A = eC. \qquad (12.20)$$

We can then determine H from Eqs. (12.6) and (12.11). We have

$$H^2 + (Z\alpha)^2 C^2 + 2H(Z\alpha)C$$
$$+ [2(Z\alpha)^2 AC + 2H(Z\alpha)A] \cos(\gamma\theta) + (Z\alpha)^2 A^2 \cos^2(\gamma\theta)$$
$$= m^2 + n_\theta^2 \left[A^2 \left(1 - \frac{(Z\alpha)^2}{n_\theta^2} \right) + C^2 + 2AC \cos(\gamma\theta) \right.$$
$$\left. + \frac{A^2(Z\alpha)^2}{n_\theta^2} \cos^2(\gamma\theta) \right]. \tag{12.21}$$

The coefficients of the terms quadratic in $\cos(\gamma\theta)$ agree, they are $(Z\alpha)^2 A^2$. The coefficients of the terms linear in $\cos(\gamma\theta)$ agree, because

$$HZ\alpha = n_\theta^2 \left(1 - \frac{(Z\alpha)^2}{n_\theta^2} \right) C = n_\theta^2 \gamma^2 C. \tag{12.22}$$

The constant terms yield

$$H^2 \left[1 + \frac{(Z\alpha)^4}{n_\theta^4 \gamma^4} + \frac{2(Z\alpha)^2}{n_\theta^2 \gamma^2} \right] = m^2 + H^2 \frac{(Z\alpha)^2}{n_\theta^2 \gamma^4} \left[\gamma^2 e^2 + 1 \right], \tag{12.23}$$

or

$$1 + \frac{(Z\alpha)^2}{n_\theta^2 \gamma^4} [(1 - \gamma^2) + 2\gamma^2 - \gamma^2 e^2 - 1] = \left(\frac{m}{H} \right)^2. \tag{12.24}$$

This simplifies to

$$1 + \frac{(Z\alpha)^2}{n_\theta^2 \gamma^4} \gamma^2 (1 - e^2) = \left(\frac{m}{H} \right)^2, \tag{12.25}$$

and, inserting $(1 - e^2)$ from Eq. (12.18):

$$1 + \frac{(Z\alpha)^2}{(n_\theta \gamma + n_r)^2} = \left(\frac{m}{H} \right)^2. \tag{12.26}$$

Equivalently, for orbits for which $H = E = constant$,

$$E = m \left(1 + \frac{(Z\alpha)^2}{(n_r + \gamma n_\theta)^2} \right)^{-\frac{1}{2}}. \tag{12.27}$$

This is the Sommerfeld formula for the energy levels of the bound states in the relativistic regime.

For the ground state, $n_\theta = 1$, the Lorentz transformation factor γ would appear in a transformation from the rest frame to a frame moving, at a given instant, in a direction tangent to the orbit with a velocity $v = Z\alpha$. Such a Lorentz transformation gives rise to a time dilation between the fixed and transformed systems by the factor

$[1 - v^2]^{-\frac{1}{2}} = [1 - (Z\alpha)^2]^{-\frac{1}{2}}$. This is the factor denoted by $1/\gamma$ in Eq. (12.8), with $p_\theta = n_\theta = 1$. The deviation of γ from one is a measure of the magnitude of relativistic effects, with $1/\gamma \geq 1$. Thus the angular velocity $\dot\theta$ appears to be smaller (from the time dilation), or, equivalently, the angle ψ in the moving system appears at each instant to be smaller, that is $\psi = \gamma\theta$.

It turns out that the quantum version of this Lorentz transformation, when applied to the operator Γ (see Chapter 15), yields the diagonalized form with the eigenvalues $\pm\gamma|\kappa|$. This is the basis of Biedenharn's analysis of the Dirac equation [Bie62], and is another hint as to why the Sommerfeld derivation yields the correct spectrum.

Not all possible transitions between the energy levels in the spectrum actually occur, they have to obey certain *selection rules*, e.g.,

$$n_{\theta_i} - n_{\theta_f} = \pm 1, \qquad (12.28)$$

where n_{θ_i} is the azimuthal quantum number in the initial state, and n_{θ_f} in the final state. The "Old Quantum Theory" cannot explain these selection rules, because it makes no predictions about the probability of transitions, or about the *intensities* of spectral lines. Neither can it explain the spectra of more complex atoms: It fails even for the next more complicated atom, namely helium with two electrons.

To get these aspects it is necessary to use the complete quantum theory. We thus turn to the Dirac theory in the next chapter.

Chapter 13
The Dirac Equation

We are now ready to tackle the relativistic problem of an electron moving rapidly in a Coulomb potential, using the full power of modern quantum mechanics. The appropriate formalism was developed by Dirac [Dir28]. In the present chapter we review the standard approach to this problem: The separation of variables in spherical polar coordinates and the derivation of differential equations for the radial components of the wave function. We emphasize that the treatment is a direct generalization of the treatment of a non-relativistic spin-$\frac{1}{2}$ particle, with the three-dimensional Clifford algebra replaced by the four-dimensional Clifford algebra, corresponding to a replacement of the two-component Pauli spinors by four-component Dirac spinors ([BJ53], [FK05]).

13.1 The Hamiltonian

In the same way as the momentum of a free non-relativistic spin-$\frac{1}{2}$ particle is described by the expression $(\boldsymbol{\sigma} \cdot \mathbf{p})$, the momentum of a free relativistic spin-$\frac{1}{2}$ particle is described by $\gamma^\mu p_\mu$, $\mu = 0, 1, 2, 3$. The γ^μ are 4×4 matrices which satisfy

$$\{\gamma^\mu, \gamma^\nu\} = \gamma^\mu \gamma^\nu + \gamma^\nu \gamma^\mu = 2g^{\mu\nu}, \tag{13.1}$$

with $g^{\mu\nu}$ the Minkowsky metric: $g^{\mu\nu} = \mathrm{diag}(1, -1, -1, -1)$. The γ^μ are the generators of a four-dimensional Clifford algebra in Minkowski space, in the same way as the σ_i are the generators of a three-dimensional Clifford algebra in Euclidian space, $\{\sigma_i, \sigma_j\} = 2\delta_{ij}$.

The equation of motion for a free particle with mass m is given by

$$\gamma^\mu p_\mu \psi(t, \mathbf{r}) = \pm m \psi(t, \mathbf{r}), \tag{13.2}$$

as is readily seen by applying $\gamma_\mu p_\mu$ twice, and using

$$\gamma^\mu \gamma^\nu p_\mu p_\nu = \frac{1}{2}\{\gamma^\mu, \gamma^\nu\} p_\mu p_\nu = p^2 = m^2, \tag{13.3}$$

which is the relativistic energy-momentum relation. With $p_\mu = i\partial_\mu$, and choosing the plus sign, yields the Dirac equation:

$$(i\gamma^\mu \partial_\mu - m)\,\psi(t, \boldsymbol{r}) = 0, \tag{13.4}$$

where we identify the parameter m with the mass. Choosing the minus sign would yield

$$(i\gamma^\mu \partial_\mu + m)\,\psi(t, \boldsymbol{r}) = 0, \tag{13.5}$$

which is the Dirac equation for a particle with negative mass $-m$. We shall consider the Eq. (13.5) to be unphysical. It will nevertheless be used in connection with the *unphysical Hamiltonian $\bar{H}$* of Eq. (15.1) below.

The gamma matrices are given (in the Dirac representation) by

$$\gamma^0 = \begin{pmatrix} I_2 & 0 \\ 0 & -I_2 \end{pmatrix}, \quad \boldsymbol{\gamma} = \begin{pmatrix} 0 & \boldsymbol{\sigma} \\ -\boldsymbol{\sigma} & 0 \end{pmatrix}, \tag{13.6}$$

with I_2 the 2×2 unity matrix, and $\boldsymbol{\sigma}$ the Pauli matrices. We shall also need the matrix

$$\gamma_5 = i\gamma^0 \gamma^1 \gamma^2 \gamma^3 = \begin{pmatrix} 0 & I_2 \\ I_2 & 0 \end{pmatrix}, \tag{13.7}$$

with $\gamma_5^2 = I_4$. This matrix anticommutes with the others: $\{\gamma_5, \gamma^\mu\} = 0$, $\mu = 0, \ldots, 3$.

The Dirac equation is cast into a Hamiltonian form by multiplying by γ^0 and introducing the new matrices

$$\beta = \gamma^0, \quad \boldsymbol{\alpha} = \gamma^0 \boldsymbol{\gamma} = \begin{pmatrix} 0 & \boldsymbol{\sigma} \\ \boldsymbol{\sigma} & 0 \end{pmatrix}. \tag{13.8}$$

Note that the new matrices are Hermitian. They satisfy

$$\{\alpha_i, \alpha_j\} = \alpha_i \alpha_j + \alpha_j \alpha_i = 2\delta_{ij}, \quad \{\alpha_i, \beta\} = 0, \tag{13.9}$$

with $i, j = 1, 2, 3$, and

$$(\alpha_i)^2 = \beta^2 = I_4. \tag{13.10}$$

The Dirac equation becomes, upon multiplying from the left by β:

$$i\frac{\partial\psi}{\partial t} = [-i\boldsymbol{\alpha}\cdot\boldsymbol{\nabla} + \beta m]\,\psi(t,\boldsymbol{r}). \tag{13.11}$$

Hence we get the Hamiltonian equation

$$i\frac{\partial\psi}{\partial t} = H\psi, \tag{13.12}$$

with

$$H = -i\boldsymbol{\alpha}\cdot\boldsymbol{\nabla} + \beta m. \tag{13.13}$$

For the stationary states $\psi(t,\mathbf{r}) = \psi(\mathbf{r})e^{-iEt}$, where E is the energy of the state, and Eq. (13.12) reduces to

$$H\psi(\mathbf{r}) = E\psi(\mathbf{r}). \tag{13.14}$$

Including a spherically symmetric potential the Hamiltonian becomes

$$H = \boldsymbol{\alpha}\cdot\mathbf{p} + \beta m + V(r), \tag{13.15}$$

with

$$\mathbf{p} = -i\boldsymbol{\nabla}. \tag{13.16}$$

13.2 Total Angular Momentum

The spin matrices are

$$\Sigma = \begin{pmatrix} \sigma & 0 \\ 0 & \sigma \end{pmatrix}. \tag{13.17}$$

We have the relation between the spin matrices and the quantities $\boldsymbol{\alpha}$:

$$\boldsymbol{\alpha} = \gamma_5\Sigma = \Sigma\gamma_5. \tag{13.18}$$

Neither the orbital angular momentum $\mathbf{L} = \mathbf{r} \times \mathbf{p}$,

$$[H, \mathbf{L}_i] = [\boldsymbol{\alpha}\cdot\mathbf{p}, \mathbf{L}_i] = \alpha_j[\mathbf{p}_j, \mathbf{L}_i] = -i\epsilon_{ijk}\alpha_j\mathbf{p}_k, \tag{13.19}$$

nor the spin vector are conserved:

$$[H, \Sigma_i] = [\alpha_j\mathbf{p}_j, \Sigma_i] = \gamma_5[\Sigma_j, \Sigma_i]\mathbf{p}_j = 2i\epsilon_{jik}\gamma_5\Sigma_k\mathbf{p}_j = 2i\epsilon_{jik}\alpha_k\mathbf{p}_j, \tag{13.20}$$

but the total angular momentum $\boldsymbol{J} = \boldsymbol{L} + \frac{1}{2}\boldsymbol{\Sigma}$ is conserved. We have used here

$$[\mathbf{L}_i, \mathbf{p}_j] = i\epsilon_{ijk}\mathbf{p}_k, \tag{13.21}$$

which expresses the vector transformation property of $\mathbf{p}$ with respect to spatial rotations. Further, relating $\boldsymbol{\alpha}$ to $\boldsymbol{\Sigma}$ by Eq. (13.18), we can use the commutator

$$[\Sigma_i, \Sigma_j] = 2i\epsilon_{ijk}\Sigma_k, \tag{13.22}$$

which expresses the so(3) structure of the Pauli matrices.

13.3 The Dirac Operator

We consider next the Dirac operator, introduced by Dirac [Dir28]

$$K = -\beta(\boldsymbol{\Sigma}\cdot\mathbf{L} + 1) = \begin{pmatrix} -(\boldsymbol{\sigma}\cdot\mathbf{L} + 1) & 0 \\ 0 & \boldsymbol{\sigma}\cdot\mathbf{L} + 1 \end{pmatrix}. \tag{13.23}$$

It is Hermitian. In the following we shall use the identity

$$(\boldsymbol{\Sigma}\cdot\mathbf{A})(\boldsymbol{\Sigma}\cdot\mathbf{B}) = \mathbf{A}\cdot\mathbf{B} + i\boldsymbol{\Sigma}\cdot(\mathbf{A} \times \mathbf{B}), \tag{13.24}$$

which holds for any two vectors $\mathbf{A}$ and $\mathbf{B}$ which commute with $\boldsymbol{\Sigma}$. This is a generalization of the identity (10.15).

The Dirac operator is a constant of the motion. Indeed, it commutes with the Hamiltonian for an arbitrary central potential V(r).

$$\begin{aligned} [H, K] &= -[K, H] = [\beta(\boldsymbol{\Sigma}\cdot\mathbf{L} + 1), \boldsymbol{\alpha}\cdot\mathbf{p} + \beta m + V(r)] \\ &= [\beta(\boldsymbol{\Sigma}\cdot\mathbf{L}), \boldsymbol{\alpha}\cdot\mathbf{p}] + [\beta, \boldsymbol{\alpha}\cdot\mathbf{p}] \\ &= \beta(\boldsymbol{\Sigma}\cdot\mathbf{L})(\boldsymbol{\alpha}\cdot\mathbf{p}) - (\boldsymbol{\alpha}\cdot\mathbf{p})\beta(\boldsymbol{\Sigma}\cdot\mathbf{L}) + 2\beta(\boldsymbol{\alpha}\cdot\mathbf{p}) \\ &= \beta(\boldsymbol{\Sigma}\cdot\mathbf{L})(\boldsymbol{\alpha}\cdot\mathbf{p}) + \beta(\boldsymbol{\alpha}\cdot\mathbf{p})(\boldsymbol{\Sigma}\cdot\mathbf{L}) + 2\beta(\boldsymbol{\alpha}\cdot\mathbf{p}) \\ &= \beta\gamma_5[(\boldsymbol{\Sigma}\cdot\mathbf{L})(\boldsymbol{\Sigma}\cdot\mathbf{p}) + (\boldsymbol{\Sigma}\cdot\mathbf{p})(\boldsymbol{\Sigma}\cdot\mathbf{L}) + 2(\boldsymbol{\Sigma}\cdot\mathbf{p})] \\ &= \beta\gamma_5[i\boldsymbol{\Sigma}\cdot(\mathbf{L} \times \mathbf{p} + \mathbf{p} \times \mathbf{L}) + 2\boldsymbol{\Sigma}\cdot\mathbf{p}], \end{aligned} \tag{13.25}$$

using the identity (13.24). Now use $\mathbf{L} \times \mathbf{p} + \mathbf{p} \times \mathbf{L} = 2i\mathbf{p}$ to get

$$[H, K] = \beta\gamma_5[-2\boldsymbol{\Sigma}\cdot\mathbf{p} + 2\boldsymbol{\Sigma}\cdot\mathbf{p}] = 0. \tag{13.26}$$

The total angular momentum commutes with the Dirac operator:

$$[\mathbf{J}, K] = -[\mathbf{J}, \beta(\boldsymbol{\Sigma}\cdot\mathbf{L} + 1)] = -\beta[\mathbf{J}, \boldsymbol{\Sigma}\cdot\mathbf{L}] = 0. \tag{13.27}$$

There is a relation between the orbital angular momentum squared and the Dirac operator squared:

$$L^2 = (\mathbf{\Sigma}\cdot\mathbf{L} + 1)(\mathbf{\Sigma}\cdot\mathbf{L}) = K(K + \beta). \tag{13.28}$$

This is the generalization of the non-relativistic relation of Eq. (10.28). There is also a relation between the total angular momentum squared and the Dirac operator squared:

$$\mathbf{J}^2 = \left(\mathbf{L} + \frac{1}{2}\mathbf{\Sigma}\right)^2 = \mathbf{L}^2 + \mathbf{\Sigma}\cdot\mathbf{L} + \frac{3}{4}$$

$$= \left(\mathbf{\Sigma}\cdot\mathbf{L} + \frac{3}{2}\right)\left(\mathbf{\Sigma}\cdot\mathbf{L} + \frac{1}{2}\right) = \left(\beta K + \frac{1}{2}\right)\left(\beta K - \frac{1}{2}\right), \tag{13.29}$$

so

$$\mathbf{J}^2 = K^2 - \frac{1}{4}. \tag{13.30}$$

This is the relativistic generalization of Eq. (10.29).

Let the eigenvalues of K be κ. Applied to a function for which K is a good quantum number, the relation (13.30) yields

$$\kappa^2 = j(j+1) + \frac{1}{4} = \left(j + \frac{1}{2}\right)^2, \tag{13.31}$$

and κ takes on two values for definite j,

$$\kappa = \pm\left|j + \frac{1}{2}\right|, \tag{13.32}$$

which generalizes Eq. (10.31).

The parity in the Dirac theory is

$$P = \beta I, \tag{13.33}$$

where I is the inversion operator which inverts the position and momenta,

$$I(\mathbf{r}) = -\mathbf{r}, \quad I(\mathbf{p}) = -\mathbf{p}. \tag{13.34}$$

It is Hermitian, and it commutes with the Hamiltonian:

$$[H, P] = [H, \beta I] = H(\mathbf{p})\beta I - \beta I H(\mathbf{p}) = \beta H(-\mathbf{p})I - \beta H(-\mathbf{p})I = 0. \tag{13.35}$$

The Dirac operator commutes with the parity:

$$[K, P] = -[\beta(\mathbf{\Sigma}\cdot\mathbf{L} + 1), \beta I]$$

$$= -\beta[\mathbf{\Sigma}\cdot\mathbf{L}, \beta I] + [\beta, \beta I](\mathbf{\Sigma}\cdot\mathbf{L} + 1) = -\beta^2[\mathbf{\Sigma}\cdot\mathbf{L}, I] = 0. \tag{13.36}$$

The total angular momentum also commutes with the parity operation:

$$[\mathbf{J}, P] = [\mathbf{J}, \beta I] = \beta[\mathbf{J}, I] = 0. \tag{13.37}$$

13.4 A Complete Set of Mutually Commuting Operators

We now have a complete set of mutually commuting operators:

$$\{H, \mathbf{J}^2, J_3, K\}. \tag{13.38}$$

To specify the parity of the states is superfluous, since its eigenvalues are known when those of $\{H, \mathbf{J}^2, J_3, K\}$ are given. We denote the eigenvectors, which we assume are normalized to unity, by

$$|\epsilon, j, m_j, \kappa\rangle, \tag{13.39}$$

where ϵ is the reduced energy

$$\epsilon = \frac{E}{m}, \tag{13.40}$$

and the quantum numbers j, m_j, κ are defined by

$$
\begin{aligned}
J^2|\epsilon, j, m_j, \kappa\rangle &= j(j+1)|\epsilon, j, m_j, \kappa\rangle, \\
J_3|\epsilon, j, m_j, \kappa\rangle &= m_j|\epsilon, j, m_j, \kappa\rangle, \\
K|\epsilon, j, m_j, \kappa\rangle &= \kappa|\epsilon, j, m_j, \kappa\rangle.
\end{aligned}
\tag{13.41}
$$

The wave function is the representation of these states in the position representation:
$\psi(\mathbf{r}) = \langle \mathbf{r}|\epsilon, j, m_j, \kappa\rangle$.

We now write the four-component wave function in terms of the two upper components and the two lower ones:

$$\psi(\mathbf{r}) = \begin{pmatrix} \psi_A(\mathbf{r}) \\ \psi_B(\mathbf{r}) \end{pmatrix}. \tag{13.42}$$

Applying the Dirac operator yields for the upper components

$$\kappa = -\boldsymbol{\sigma} \cdot \mathbf{L} - 1, \tag{13.43}$$

and using $\mathbf{J} = \mathbf{L} + \frac{1}{2}\boldsymbol{\sigma}$ we find

$$\mathbf{J}^2 = \mathbf{L}^2 + \boldsymbol{\sigma} \cdot \mathbf{L} + \frac{3}{4}, \tag{13.44}$$

or

$$\boldsymbol{\sigma} \cdot \mathbf{L} = \mathbf{J}^2 - \mathbf{L}^2 - \frac{3}{4}, \tag{13.45}$$

and Eq. (13.43) becomes

$$\kappa = \mathbf{L}^2 - \mathbf{J}^2 - \frac{1}{4}. \tag{13.46}$$

If $\boldsymbol{\sigma} \cdot \mathbf{L}$ is a good quantum number so is $\mathbf{L}^2$, from Eq. (13.45). So

$$\kappa = l_A(l_A + 1) - j(j + 1) - \frac{1}{4}. \tag{13.47}$$

For $l_A = j + \frac{1}{2}$ we find $\kappa = l_A$, for $l_A = j - \frac{1}{2}$ we find $\kappa = -l_A - 1$. By vector addition of angular momentum these are the only two possibilities. The result is in agreement with Eq. (13.32). We see that $l_A = j + \frac{1}{2}Sign(\kappa)$, with $Sign(\kappa) = \kappa/|\kappa|$.

For the lower components $\kappa = \boldsymbol{\sigma} \cdot \mathbf{L} + 1$, and $\kappa = l_B + 1$ for $l_B = j - \frac{1}{2}$, $\kappa = -l_B$ for $l_B = j + \frac{1}{2}$, and $l_B = j - \frac{1}{2}Sign(\kappa)$. We have $l_A - l_B = Sign(\kappa)$. The cases where $j = l_{A,B} + 1/2$ are referred to as "parallel coupling", the cases where $j = l_{A,B} - 1/2$ as "antiparallel coupling".

The spatial parity of $\psi_A(\mathbf{r})$ is $(-1)^{l_A}$, the spatial parity of $\psi_B(\mathbf{x})$ is $(-1)^{l_B} = -(-1)^{l_A}$. But the conserved quantity is the total parity $P = \beta I$. The total parity of $\psi(\mathbf{r})$ is $(-1)^{l_A} = -(-1)^{l_B} = (-1)^{j+\frac{1}{2}Sign(\kappa)}$.

13.5 The Dirac Spinors

We can now write the Dirac spinors in the form

$$\psi(\epsilon, j, m_j, \kappa; \mathbf{r}) = \begin{pmatrix} \psi_A(\mathbf{r}) \\ \psi_B(\mathbf{r}) \end{pmatrix} = \begin{pmatrix} R_A(r)\mathcal{Y}_{jl_A}^{m_j}(\mathbf{r}) \\ iR_B(r)\mathcal{Y}_{jl_B}^{m_j}(\hat{\mathbf{r}}) \end{pmatrix}, \tag{13.48}$$

where $\mathcal{Y}_{jl}^{j_3}(\hat{\mathbf{r}})$ is a normalized spin-angular function formed by the combination of the Pauli spinors with the spherical harmonics of order l. The radial functions R_A and R_B depend on κ. The factor i is inserted to make R_A and R_B real for bound-state solutions.

13.6 The Radial Equations in Polar Coordinates

We rewrite the Hamiltonian in spherical polar coordinates. We begin with the term $\boldsymbol{\Sigma} \cdot \mathbf{p}$:

$$\boldsymbol{\Sigma} \cdot \mathbf{p} = (\boldsymbol{\Sigma} \cdot \hat{\mathbf{r}})(\boldsymbol{\Sigma} \cdot \hat{\mathbf{r}})\boldsymbol{\Sigma} \cdot \mathbf{p} = (\boldsymbol{\Sigma} \cdot \hat{\mathbf{r}})(\hat{\mathbf{r}} \cdot \mathbf{p} + i\boldsymbol{\Sigma} \cdot (\hat{\mathbf{r}} \times \mathbf{p})) = (\boldsymbol{\Sigma} \cdot \hat{\mathbf{r}}) \left(\hat{\mathbf{r}} \cdot \mathbf{p} + \frac{i}{r}\boldsymbol{\Sigma} \cdot \mathbf{L} \right), \tag{13.49}$$

so

$$\mathbf{\Sigma \cdot p} = i(\mathbf{\Sigma \cdot \hat{r}})\left(-\frac{\partial}{\partial r} + \frac{\mathbf{\Sigma \cdot L}}{r}\right). \tag{13.50}$$

Now rewrite $\mathbf{\Sigma \cdot L}$ in terms of the Dirac operator K,

$$\mathbf{\Sigma \cdot p} = -i(\mathbf{\Sigma \cdot \hat{r}})\left(\frac{\partial}{\partial r} + \frac{\beta K + 1}{r}\right). \tag{13.51}$$

This is the relativistic generalization of Eq. (10.66). Multiply this equation from the left by γ_5 and get, by Eq. (13.18):

$$\mathbf{\alpha \cdot p} = -i(\mathbf{\alpha \cdot \hat{r}})\left(\frac{\partial}{\partial r} + \frac{\beta K + 1}{r}\right). \tag{13.52}$$

We obtain for the Hamiltonian the expression

$$H = -i(\mathbf{\alpha \cdot \hat{r}})\left(\frac{\partial}{\partial r} + \frac{\beta K + 1}{r}\right) + \beta m + V(r). \tag{13.53}$$

With this Hamiltonian the equation for the bound states becomes

$$\left(i\begin{pmatrix}0 & \mathbf{\sigma \cdot \hat{r}} \\ \mathbf{\sigma \cdot \hat{r}} & 0\end{pmatrix}\left(-\frac{\partial}{\partial r} - \frac{1}{r}\right) - i\begin{pmatrix}0 & -\mathbf{\sigma \cdot \hat{r}} \\ \mathbf{\sigma \cdot \hat{r}} & 0\end{pmatrix}\frac{\kappa}{r} + \begin{pmatrix}m & 0 \\ 0 & -m\end{pmatrix} + V(r)\right)$$

$$\times \begin{pmatrix}R_A \mathcal{Y}_A \\ iR_B \mathcal{Y}_B\end{pmatrix} = E\begin{pmatrix}R_A \mathcal{Y}_A \\ iR_B \mathcal{Y}_B\end{pmatrix}. \tag{13.54}$$

In terms of the two-component spinors the upper part of Eq. (13.54) is

$$\left(\frac{dR_B}{dr} + \frac{(1 - \kappa)R_B}{r}\right)(\mathbf{\sigma \cdot \hat{r}})\mathcal{Y}_B = (E - m - V(r))\,R_A \mathcal{Y}_A. \tag{13.55}$$

Now use Eq. (10.46) and get

$$\left(-\frac{dR_B}{dr} + \frac{(\kappa - 1)R_B}{r}\right)\mathcal{Y}_A = (E - m - V(r))\,R_A \mathcal{Y}_A, \tag{13.56}$$

or

$$-\frac{dR_B}{dr} + \frac{(\kappa - 1)}{r}R_B = (E - m - V(r))\,R_A. \tag{13.57}$$

Similarly, the lower part of Eq. (13.54) is

$$\frac{dR_A}{dr} + \frac{(\kappa + 1)}{r}R_A = (E + m - V(r))\,R_B. \tag{13.58}$$

We introduce the functions

$$u_A(r) = rR_A(r), \quad u_B(r) = rR_B(r), \tag{13.59}$$

and the dimensionless variable x defined by

$$x = 2m\sqrt{1 - \epsilon^2}\, r = \left(\frac{2Z\alpha m}{n_a}\right) r , \tag{13.60}$$

where

$$n_a = \frac{Z\alpha}{\sqrt{1 - \epsilon^2}}. \tag{13.61}$$

Since in the non-relativistic limit

$$\epsilon \to 1 - \frac{1}{2}\frac{(Z\alpha)^2}{N^2} + O((Z\alpha)^4) \tag{13.62}$$

we have

$$n_a \to N \tag{13.63}$$

in this limit. For this reason n_a is called the "apparent principal quantum number".

In terms of the functions $u_A(x), u_B(x)$ we find the coupled differential equations

$$\left(\frac{d}{dx} - \frac{\kappa}{x}\right) u_B(x) = \left(\frac{1}{2}\sqrt{\frac{1 - \epsilon}{1 + \epsilon}} - \frac{Z\alpha}{x}\right) u_A(x),$$
$$\left(\frac{d}{dx} + \frac{\kappa}{x}\right) u_A(x) = \left(\frac{1}{2}\sqrt{\frac{1 + \epsilon}{1 - \epsilon}} + \frac{Z\alpha}{x}\right) u_B(x). \tag{13.64}$$

Equation (13.64) may be written in the following matrix form:

$$\left(\frac{du_A}{dx}, \frac{du_B}{dx}\right) = (u_a, u_B)\frac{1}{x}\begin{pmatrix} -\kappa & -Z\alpha \\ Z\alpha & \kappa \end{pmatrix}$$
$$+ (u_A, u_B)\frac{1}{2\sqrt{1 - \epsilon^2}}\begin{pmatrix} 0 & 1 - \epsilon \\ 1 + \epsilon & 0 \end{pmatrix}. \tag{13.65}$$

The solution of these coupled differential equations in the literature is somewhat *ad hoc*. The recognition of the supersymmetric structure of the equations allows a systematic development.

Notes on Chapter 13

This is the work of Dirac [Dir28], referred to in Chapter 1. The considerations here are by now completely standard; see the texts by Bjorken and Drell [BD64], Sakurai [Sak67], or Rose [Ros61]. The Dirac equation is here motivated as a generalization of the Pauli equation in Chapter 10.

Chapter 14

The Primary Supersymmetry of the Dirac Equation

An important conserved quantity of the Dirac equation was introduced by Johnson and Lippmann [JL50], designated here as A. As we shall see at the end of this chapter, it is the relativistic generalization of $A_{nr} = (\boldsymbol{\sigma} \cdot \mathcal{A})$, where $\mathcal{A}$ is the Laplace vector. It is only announced in the original paper; the first derivation was given by Biedenharn [Bie62]. The present derivation is by Khachidze and Khelashvili [KK06].

We show that, acting on an eigenstate of the Dirac operator K, A has the effect of interchanging the eigenvalues κ and $-\kappa$. The equation

$$A|\epsilon, j, m_j, \kappa > = -e|\epsilon, j, m_j, -\kappa > \tag{14.1}$$

thus replaces the key equation of our non-relativistic analysis of the spectrum of the hydrogen atom. This equation expresses the *primary supersymmetry* of the Dirac equation. The supersymmetry of the Dirac equation in a Coulomb field may be seen just by looking at the pattern formed by the lines of the spectrum in Figure 14.1.

14.1 A Derivation of the Johnson–Lippmann Operator

We first generalize the theorem of Section 10.2 to the relativistic context.

Theorem. For any vector $\mathbf{v}$ with respect to L, which is perpendicular to L,

$$\mathbf{v} \cdot \mathbf{L} = \mathbf{L} \cdot \mathbf{v} = 0, \tag{14.2}$$

the Dirac operator K anticommutes with $(\boldsymbol{\Sigma} \cdot \mathbf{v})$.

Proof. The proof is a straightforward generalization of the previous case. The corollary is

$$K(\boldsymbol{\Sigma} \cdot \mathbf{v}) = (i/2)\beta \boldsymbol{\Sigma} \cdot (\mathbf{v} \times \mathbf{L} - \mathbf{L} \times \mathbf{v}). \tag{14.3}$$

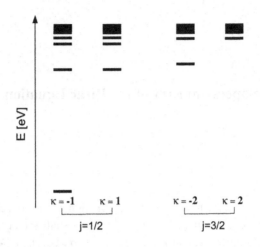

Figure 14.1.   The supersymmetry of the relativistic hydrogen spectrum.

We again have as examples $\mathbf{v} = \hat{r}$, $\mathbf{v} = \mathbf{p}$, $\mathbf{v} = \mathcal{A}$, where, by Eq. (8.8),

$$\mathcal{A} = \frac{1}{2Z\alpha m}(\mathbf{p} \times \mathbf{L} - \mathbf{L} \times \mathbf{p}) - \hat{r}. \qquad (14.4)$$

There is a relation between these three K-odd operators:

$$\boldsymbol{\Sigma}\cdot\mathcal{A} = \frac{1}{2Z\alpha m}\boldsymbol{\Sigma}\cdot(\mathbf{p} \times \mathbf{L} - \mathbf{L} \times \mathbf{p}) - \boldsymbol{\Sigma}\cdot\hat{r} = \frac{-i}{Z\alpha m}\beta K(\boldsymbol{\Sigma}\cdot\mathbf{p}) - \boldsymbol{\Sigma}\cdot\hat{r},$$
$$(14.5)$$

where we have used Eq. (14.3).

The class of operators that anticommute with K, the K-odd operators, is not restricted to the operators just discussed. Any operator of the form $O(\boldsymbol{\Sigma}\cdot\mathbf{v})$, where O is any operator which commutes with K, is also K-odd.

In the non-relativistic case, we found the key equation, Eq. (10.57), by utilizing a constant of the motion which is K-odd. We try the same strategy here: We look for a K-odd constant of the motion. Let us try a combination of $(\boldsymbol{\Sigma}\cdot\hat{r})$ and $K(\boldsymbol{\Sigma}\cdot\mathbf{p})$, where the second is of the form $O(\boldsymbol{\Sigma}\cdot\mathbf{p})$. This choice is motivated by Eq. (14.5), which relates these operators to the Laplace vector. Both of the above operators are represented by diagonal matrices. After commuting them with the non-diagonal operator H, to check if they are constants of the motion, we will get anti-diagonal terms. For example,

$$[\boldsymbol{\Sigma}\cdot\hat{r}, H] = \frac{-2i}{r}\beta K\gamma_5. \qquad (14.6)$$

Exercise 14.1 Prove Eq. (14.6).

This new operator is anti-diagonal. It is only available in the relativistic case. For this reason we consider the following combination of diagonal and anti-diagonal operators:

$$A' = x_1(\Sigma \cdot \hat{\mathbf{r}}) + ix_2 K(\Sigma \cdot \mathbf{p}) + ix_3 K \gamma_5 f(r), \tag{14.7}$$

where $f(r)$ is an arbitrary scalar function to be determined later. The coefficients are chosen in such a way that A' is Hermitian with x_1, x_2, x_3 arbitrary real numbers. Indeed,

$$A'^\dagger = x_1(\Sigma \cdot \hat{\mathbf{r}}) - ix_2(\Sigma \cdot \mathbf{p})K - ix_3 \gamma_5 K f(r) = A'. \tag{14.8}$$

In order that the K-odd operator A' be a constant of the motion we must have $[A', H] = 0$. To calculate this commutator we need the following terms:

Exercise 14.2 Prove $[K(\Sigma \cdot \mathbf{p}), H] = -iK(\Sigma \cdot \hat{\mathbf{r}})V'(r)$.

Exercise 14.3 Prove $[K\gamma_5 f(r), H] = iK(\Sigma \cdot \hat{\mathbf{r}})f'(r) + 2mK\gamma_5 \beta f(r)$.

Together with Eq. (14.6) these yield

$$[A', H] = -\frac{2i}{r}x_1\beta K\gamma_5 + x_2 V'(r)K(\Sigma \cdot \hat{\mathbf{r}})$$
$$- x_3 f'(r)K(\Sigma \cdot \hat{\mathbf{r}}) - 2ix_3 m\beta K\gamma_5 f(r). \tag{14.9}$$

We group diagonal and anti-diagonal operators separately and set Eq. (14.9) to zero. This yields

$$-2i\beta K\gamma_5 \left[\frac{x_1}{r} + mx_3 f(r)\right] + K(\Sigma \cdot \hat{\mathbf{r}})[x_2 V'(r) - x_3 f'(r)] = 0. \tag{14.10}$$

This is fulfilled if the diagonal and the anti-diagonal terms vanish separately:

$$\frac{x_1}{r} = -x_3 m f(r), \quad x_2 V'(r) = x_3 f'(r). \tag{14.11}$$

We integrate the second equation over the interval (r, ∞), with the requirement that $f(r)$ and $V(r)$ tend to zero when $r \to \infty$, and find

$$x_2 V(r) = x_3 f(r). \tag{14.12}$$

From the first equation

$$f(r) = -\frac{x_1}{x_3}\frac{1}{mr}. \tag{14.13}$$

So

$$V(r) = -\frac{x_1}{x_2}\frac{1}{mr}. \tag{14.14}$$

Hence we have shown, within a very general framework, that the only potential for which the Dirac Hamiltonian has an additional symmetry is a *Coulomb potential*. Since the relative signs of x_1 and x_2 are arbitrary, we have a symmetry both for attraction and repulsion. For attraction

$$V(r) = -\frac{Z\alpha}{r}, \tag{14.15}$$

and it follows that

$$x_2 = \frac{x_1}{Z\alpha m}. \tag{14.16}$$

The operator (14.7) becomes

$$A' = x_1 \left[\boldsymbol{\Sigma}\cdot\hat{\mathbf{r}} + \frac{i}{Z\alpha m}K(\boldsymbol{\Sigma}\cdot\mathbf{p}) - \frac{i}{mr}K\gamma_5 \right]. \tag{14.17}$$

This K-odd constant of the motion is actually a constant multiple of the operator discovered by Johnson and Lippman [JL50], conventionally written in the form

$$A = \frac{-i}{Z\alpha m}K\gamma_5(H - \beta m) - \boldsymbol{\Sigma}\cdot\hat{\mathbf{r}}. \tag{14.18}$$

Substituting the expressions for H, Eq. (13.15), and K, Eq. (13.23), into the expression for A brings the operator into the following form:

$$A = \frac{i}{Z\alpha m}\beta(\boldsymbol{\Sigma}\cdot\mathbf{L} + 1)(\boldsymbol{\Sigma}\cdot\mathbf{p}) - \frac{iK\gamma_5}{mr} - \boldsymbol{\Sigma}\cdot\hat{\mathbf{r}}. \tag{14.19}$$

From Eq. (14.3)

$$-(\boldsymbol{\Sigma}\cdot\mathbf{L} + 1)(\boldsymbol{\Sigma}\cdot\mathbf{p}) = \frac{i}{2}\boldsymbol{\Sigma}\cdot(\mathbf{p} \times \mathbf{L} - \mathbf{L} \times \mathbf{p}), \tag{14.20}$$

and this enables us to write A as

$$A = \boldsymbol{\Sigma}\cdot\left(\frac{1}{2Z\alpha m}\beta(\mathbf{p} \times \mathbf{L} - \mathbf{L} \times \mathbf{p}) - \hat{\mathbf{r}} \right) - \frac{iK\gamma_5}{mr}. \tag{14.21}$$

The non-relativistic limit of the Johnson–Lippmann operator is obtained, in the Dirac representation of the γ–matrices, by taking the "large" components of the Dirac spinors, that is, restricting oneself to the upper two components. This amounts to replacing $\beta \to 1$ and $\gamma_5 \to 0$. This is because the effect of γ_5 on the large components is to add an admixture of the small components to them, which is a negligible effect in the non-relativistic limit. We then see that

$$A \to A_{nr} = (\boldsymbol{\sigma} \cdot \boldsymbol{\mathcal{A}}). \tag{14.22}$$

In this sense the Johnson–Lippmann operator A is the relativistic generalization of $(\boldsymbol{\sigma} \cdot \boldsymbol{\mathcal{A}})$, with $\boldsymbol{\mathcal{A}}$ the Laplace vector.

14.2 Commutation and Anticommutation Relations of the Johnson–Lippmann Operator

The Johnson–Lippmann operator commutes with the total angular momentum **J**:

$$[\mathbf{J}, A] = -\frac{i}{Z\alpha m}[\mathbf{J}, K\gamma_5(H - \beta m)] - \left[\mathbf{J}, \frac{\boldsymbol{\Sigma} \cdot \mathbf{r}}{r}\right] = 0. \tag{14.23}$$

This is because each of the commutators vanish separately; the first because J commutes with each of the factors, the second because any scalar product and any function of r is invariant with respect to rotations.

The Johnson–Lippmann operator anticommutes with the Dirac operator K. K anticommutes with the first term of A: Since K commutes with the factor $(H - \beta m)$ it suffices to show that K anticommutes with $K\gamma_5$:

$$K(K\gamma_5) = -K(\gamma_5 K) = -(K\gamma_5)K. \tag{14.24}$$

Hence K anticommutes with the complete expression for the Johnson–Lippmann operator A,

$$AK + KA = 0. \tag{14.25}$$

The Johnson–Lippmann operator A also anticommutes with the total parity operator $P = \beta I$. Write A in the form (14.18). K commutes with $(H - \beta m)$. So the first term in A anticommutes with P:

$$K\gamma_5\beta I = -\beta(\boldsymbol{\Sigma} \cdot \mathbf{L} + 1)\gamma_5\beta I = \beta(\boldsymbol{\Sigma} \cdot \mathbf{L} + 1)\beta\gamma_5 I$$

$$= \beta I(\boldsymbol{\Sigma} \cdot \mathbf{L} + 1)\beta\gamma_5 = \beta I\beta(\boldsymbol{\Sigma} \cdot \mathbf{L} + 1)\gamma_5$$

$$= -\beta I K\gamma_5. \tag{14.26}$$

Also the second term in A anticommutes with P:

$$\boldsymbol{\Sigma}\cdot\mathbf{r}\beta I = \beta\boldsymbol{\Sigma}\cdot\mathbf{r}I = -\beta I\boldsymbol{\Sigma}\cdot\mathbf{r}. \tag{14.27}$$

So the parity operator P anticommutes with the complete expression for the Johnson–Lippmann operator A,

$$AP + PA = 0. \tag{14.28}$$

14.3 Eccentricity

We now derive an expression for the magnitude of the Johnson–Lippmann operator in terms of the eigenvalues of the Dirac operator. We evaluate the operator A^2 as $A^\dagger A$, and derive the following expression for A^2:

$$A^2 = \left[\frac{i}{Z\alpha m}(H - \beta m)\gamma_5 K - \boldsymbol{\Sigma}\cdot\hat{\mathbf{r}}\right]\left[\frac{-i}{Z\alpha m}K\gamma_5(H - \beta m) - \boldsymbol{\Sigma}\cdot\hat{\mathbf{r}}\right]$$

$$= \frac{1}{(Z\alpha m)^2}K^2(H - \beta m)^2 + 1$$

$$- \frac{i}{Z\alpha m}[(\boldsymbol{\Sigma}\cdot\hat{\mathbf{r}})K\gamma_5(H - \beta m) - (H - \beta m)\gamma_5 K(\boldsymbol{\Sigma}\cdot\hat{\mathbf{r}})]. \tag{14.29}$$

Exercise 14.4 Prove $K^2(H - \beta m)^2 = K^2\left(H^2 - m^2 + 2m\beta\frac{Z\alpha}{r}\right)$.

Exercise 14.5 Prove $(\boldsymbol{\Sigma}\cdot\hat{r})K\gamma_5(H-\beta m) -(H-\beta m)\gamma_5 K(\boldsymbol{\Sigma}\cdot\hat{r}) = -2i\beta K^2/r$.

Putting the two pieces together, and adding 1:

$$A^2 = 1 + \frac{K^2}{(Z\alpha m)^2}\left(H^2 - m^2 + 2m\beta\frac{Z\alpha}{r}\right) - \frac{i}{Z\alpha m}(-2i)\frac{\beta K^2}{r}$$

$$= 1 + \frac{1}{(Z\alpha m)^2}K^2(H^2 - m^2). \tag{14.30}$$

Replacing the operators in this relation with their eigenvalues gives

$$e^2 = 1 - \frac{\kappa^2}{(Z\alpha)^2}(1 - \epsilon^2), \tag{14.31}$$

where e is the magnitude of A. It is convenient to rewrite this result in the form

$$e^2 = 1 - \left(\frac{\kappa}{n_a}\right)^2, \tag{14.32}$$

where n_a is defined by the Eq. (13.61). We have defined e to be non-negative; hence we have

$$e = \sqrt{1 - \left(\frac{\kappa}{n_a}\right)^2}. \tag{14.33}$$

e will be interpreted as the *eccentricity*. The similarity to the expression for the non-relativistic eccentricity, Eq. (10.80), and to the classical eccentricity, Eq. (2.29), is remarkable.

14.4 The Johnson–Lippmann Operator as the Generator of Supersymmetry

The Johnson–Lippmann operator is a conserved quantity, which, when operating on an eigenstate of κ, interchanges κ and $-\kappa$. If the eigenvalue $\kappa > 0$ characterizes one sort of particle, and $\kappa < 0$ characterizes the other sort, then A is the supersymmetric charge which communicates between the two sorts. It thus generates a supersymmetric algebra. This is the *primary* supersymmetry of the Dirac equation. This was first noted by Nieto *et al.* [NKT85]. Applied analogously to the non-relativistic case it yields solutions for the radial wave functions.

The fact that A commutes with H and $\mathbf{J}$ but anticommutes with K implies first that $A|\epsilon, j, m_j, \kappa\rangle$ is zero or else an eigenstate of $H, \mathbf{J}^2$, and $\mathbf{J}_3$ with the same eigenvalues as $|\epsilon, j, m_j, \kappa\rangle$, and next that it is an eigenstate of K with an eigenvalue that is the opposite of that associated with $|\epsilon, j, m_j, \kappa\rangle$. Thus we may write

$$A|\epsilon, j, m_j, \kappa\rangle = -e|\epsilon, j, m_j, -\kappa >, \tag{14.34}$$

where e is a so-far undetermined constant. By a proper choice of the relative phase of the vector sets $|\epsilon, j, m_j, \kappa\rangle$ and $|\epsilon, j, m_j, -\kappa >$, we may ensure that e is real and positive (or zero). The fact that A is a Hermitian operator implies then that we also have

$$A|\epsilon, j, m_j, -\kappa >= -e|\epsilon, j, m_j, \kappa\rangle, \tag{14.35}$$

where e is the same as before.

We see that

$$A\left(|\epsilon, j, m_j, \kappa\rangle \pm |\epsilon, j, m_j, -\kappa\rangle\right) = \mp e\left(|\epsilon, j, m_j, \kappa\rangle \pm |\epsilon, j, m_j, -\kappa\rangle\right). \tag{14.36}$$

The state vectors $\left(|\epsilon, j, m_j, \kappa\rangle \pm |\epsilon, j, m_j, -\kappa\rangle\right)$ are, accordingly, eigenstates of the alternative set of mutually commuting operators $\{H, \mathbf{J}^2, \mathbf{J}_3, A\}$.

We have two types of states, those with positive values of κ, and those with negative values of κ, with the operator A mediating between them, Eqs. (14.34) and (14.35), so it is natural to identify

$$Q = \begin{bmatrix} 0 & 0 \\ A & 0 \end{bmatrix}, \quad Q^\dagger = \begin{bmatrix} 0 & A \\ 0 & 0 \end{bmatrix} \tag{14.37}$$

with the supersymmetry generators, i.e., we have $A^+ = A^- = A$, since A is Hermitian. The supersymmetric Hamiltonian becomes

$$\mathcal{H} = \begin{bmatrix} H_+ & 0 \\ 0 & H_- \end{bmatrix} = \begin{pmatrix} A^2 & 0 \\ 0 & A^2 \end{pmatrix}. \tag{14.38}$$

We recall that A commutes with $\mathbf{J}^2$ and $\mathbf{J}_3$, which implies that the supersymmetric generators, and the supersymmetric Hamiltonian also commute with these operators. Thus, the S(2) supersymmetry is separate from the SO(3) symmetry induced by the central field, and the total symmetry of the Dirac–Kepler problem is SO(3) × SO(2).

In accordance with this, we have a pair of bosonic and fermionic stacks for each level of j. In fact we have $2j + 1$ identical pair of stacks for each j, corresponding to the $2j + 1$ values of m_j. In the following, we shall assume that all levels of a pair of stacks have been assigned the same fixed values of the SU(2) quantum numbers j and m_j.

We identify the vectors $|\nu_1\rangle$ and $|\nu_2\rangle$ with the Dirac spinors, but suppress the quantum numbers j and m_j in accordance with the above remarks. Thus, we have

$$\psi(r; \epsilon) = \begin{bmatrix} \psi(r; \epsilon, |\kappa|) \\ \psi(r; \epsilon, -|\kappa|) \end{bmatrix}. \tag{14.39}$$

The parity operator that is responsible for the grading is

$$\mathcal{P}_\kappa = K/|\kappa|, \tag{14.40}$$

where K is the Dirac operator. It is represented by the matrix specified in Eq. (11.8).

The functions $\psi(r; \epsilon, |\kappa|)$ and $\psi(r; \epsilon, -|\kappa|)$ are transformed into each other by the operator A, and they are both eigenfunctions of A^2 with eigenvalue e^2. Thus, $\psi(r; \epsilon)$ is an eigenvector of the supersymmetric Hamiltonian. The relation (14.33) shows that there is a one-to-one correspondence between the values of ϵ and e in a given pair of stacks, i.e., for a fixed value of j and, hence, of $|\kappa|$. The supersymmetric Hamiltonian $\mathcal{H}$ is thus closely tied to the Hamiltonian H, as it should be.

Having completely identified the exact supersymmetry of the Dirac–Kepler problem, we must determine the $\psi(r; \epsilon)$ corresponding to the bottom level, for which the eigenvalue of $\mathcal{H}$, and, hence, e is zero. We ask, accordingly, for a Dirac spinor $\psi_0(r; \epsilon, |\kappa|)$ such that

$$A\psi_0(r; \epsilon, |\kappa|) = 0. \tag{14.41}$$

In the notation of Eq. (13.48), we must determine radial functions $R_A(r)$ and $R_B(r)$ satisfying the equation

$$A \begin{pmatrix} R_A(r)\mathcal{Y}_A \\ iR_B(r)\mathcal{Y}_B \end{pmatrix} = \begin{pmatrix} 0 \\ 0 \end{pmatrix}. \tag{14.42}$$

From Eq. (13.53)

$$H - \beta m = -i(\boldsymbol{\alpha} \cdot \hat{\mathbf{r}}) \left(\frac{\partial}{\partial r} + \frac{\beta K}{r} + \frac{1}{r} \right) - \frac{Z\alpha}{r}. \tag{14.43}$$

We also use the relations (10.46) and (10.55) to get

$$(\boldsymbol{\Sigma} \cdot \hat{\mathbf{r}}) \begin{pmatrix} R_A(r)\mathcal{Y}_A \\ iR_B(r)\mathcal{Y}_B \end{pmatrix} = - \begin{pmatrix} R_A(r)\mathcal{Y}_B \\ iR_B(r)\mathcal{Y}_A \end{pmatrix}, \tag{14.44}$$

which holds for any pair of radial functions. We use the matrix form of γ_5, Eq. (13.1), and the fact that $\psi_0(r; \epsilon, \kappa)$ is an eigenfunction of K with eigenvalue κ. Then, using the explicit form of A, Eq. (14.18), in Eq. (14.42):

$$\begin{aligned} A\psi_0(r; \epsilon, \kappa) &= \left(\frac{-i}{Z\alpha m} K\gamma_5 (H - \beta m) - \boldsymbol{\Sigma} \cdot \hat{\mathbf{r}} \right) \begin{pmatrix} R_A(r))\mathcal{Y}_A \\ iR_B(r)\mathcal{Y}_B \end{pmatrix} \\ &= -\frac{\kappa}{Z\alpha m} \left(\frac{\partial}{\partial r} + \frac{\beta \kappa}{r} + \frac{1}{r} \right) \begin{pmatrix} R_A(r)\mathcal{Y}_B \\ iR_B(r)\mathcal{Y}_A \end{pmatrix} \\ &\quad - \frac{\kappa}{mr} \begin{pmatrix} -R_B(r)\mathcal{Y}_B \\ iR_A(r)\mathcal{Y}_A \end{pmatrix} + \begin{pmatrix} R_A(r)\mathcal{Y}_B \\ iR_B(r)\mathcal{Y}_A \end{pmatrix} = \begin{pmatrix} 0 \\ 0 \end{pmatrix}. \end{aligned} \tag{14.45}$$

Hence, the condition (14.41) becomes

$$\left(-\frac{Z\alpha m}{\kappa}r + r\frac{d}{dr} + \kappa + 1 \right) R_A(r) = Z\alpha R_B(r),$$

$$\left(-\frac{Z\alpha m}{\kappa}r + r\frac{d}{dr} - \kappa + 1 \right) R_B(r) = -Z\alpha R_A(r). \tag{14.46}$$

Using the variable

$$x = \left(\frac{2Z\alpha m}{|\kappa|} \right) r, \tag{14.47}$$

which, as we shall see below, is consistent in this case with (13.60) (because $|\kappa| = n_a$), we get instead

$$\left(-\frac{1}{2}Sign(\kappa)x + x\frac{d}{dx} + \kappa + 1\right) R_A(x) = Z\alpha R_B(x),$$

$$\left(-\frac{1}{2}Sign(\kappa)x + x\frac{d}{dx} - \kappa + 1\right) R_B(x) = -Z\alpha R_A(x). \qquad (14.48)$$

Inserting one of these equations into the other shows that both $R_A(x)$ and $R_B(x)$ satisfy the equation

$$\left[\left(\frac{d}{dx}x - \frac{1}{2}Sign(\kappa)x\right)^2 - \gamma^2|\kappa|^2\right] R(x) = 0, \qquad (14.49)$$

with

$$\gamma^2 = 1 - \frac{(Z\alpha)^2}{|\kappa|^2}. \qquad (14.50)$$

This equation factorizes as follows:

$$\left(\frac{d}{dx}x - \frac{1}{2}Sign(\kappa)x - \gamma|\kappa|\right)\left(\frac{d}{dx}x - \frac{1}{2}Sign(\kappa)x + \gamma|\kappa|\right) R(x) = 0. \qquad (14.51)$$

The order of the two factors is arbitrary. Hence, we end up with the two equations

$$\left(\frac{d}{dx}x - \frac{1}{2}Sign(\kappa)x - \gamma|\kappa|\right) R(x) = 0,$$

$$\left(\frac{d}{dx}x - \frac{1}{2}Sign(\kappa)x + \gamma|\kappa|\right) R(x) = 0, \qquad (14.52)$$

with the respective solutions

$$R(x) = (const)x^{\gamma|\kappa|-1}e^{\frac{1}{2}xSign(\kappa)}$$

$$R(x) = (const)x^{-\gamma|\kappa|-1}e^{\frac{1}{2}xSign(\kappa)}. \qquad (14.53)$$

The second solution is physically unacceptable because of its behavior at the origin. The first solution behaves properly at the origin, but it is only normalizable for negative values of κ. Hence, we conclude that Eq. (14.41) only has a proper solution for negative values of κ. Identifying the solution $R(x)$ with $R_A(x)$ and inserting it into Eq. (14.48) gives

$$R_B(x) = \frac{\kappa + \gamma|\kappa|}{Z\alpha} R_A(x) = \frac{Z\alpha}{\kappa - \gamma|\kappa|} R_A(x). \qquad (14.54)$$

The $\psi_0(r; \epsilon)$ corresponding to the lowest energy of a double stack has now been found to have the form

$$\psi_0(r; \epsilon) = \begin{bmatrix} 0 \\ \psi_0(r; \epsilon, \kappa) \end{bmatrix}, \qquad (14.55)$$

with

$$\psi_0(r; \epsilon, \kappa) = R_0(x) \begin{pmatrix} \mathcal{Y}_A \\ i \frac{\kappa + \gamma|\kappa|}{Z\alpha} \mathcal{Y}_B \end{pmatrix}, \qquad (14.56)$$

and

$$R_0(x) = (const)x^{\gamma|\kappa|-1}e^{-\frac{x}{2}}. \qquad (14.57)$$

κ is negative and the functions $\mathcal{Y}_A$ and $\mathcal{Y}_B$ must be chosen according to the rules for parallel coupling, described in Section 13.4. Thus, we have identified the ground state of the supersymmetric Hamiltonian. The corresponding eigenvector satisfies the equations

$$Q\psi_0(r; \epsilon) = 0, \quad Q^\dagger\psi_0(r; \epsilon) = 0, \qquad (14.58)$$

as it should for the supersymmetry to be exact.

The energy of the ground state level with a fixed value of κ is easily obtained from the results of Section 14.3 with $e = 0$. The relation (14.33) shows first that $n_a = |\kappa|$, and when this is inserted in Eq. (14.31), we get

$$\epsilon = \sqrt{1 - \left(\frac{Z\alpha}{\kappa}\right)^2} = \gamma, \qquad (14.59)$$

with γ as in Eq. (14.50). We also note that the relation $n_a = |\kappa|$ implies that the definitions (13.60) and (14.47) of the variable x coincide when $e = 0$.

Such states correspond to the "circular" orbits of the Bohr model. They are non-degenerate under supersymmetry and hence correspond to the eigenvalue $e = 0$ for the operator A, which is the relativistic generalization of $A_{nr} = (\boldsymbol{\sigma} \cdot \mathcal{A})$. Thus we have established the result that the states corresponding to the circular orbits are those with vanishing eccentricity.

Having determined the ground-level wave function for a double stack corresponding to a given level of $|\kappa|$, we should focus on the energies and wave functions of the higher levels. This could be done by solving the equation

$$\mathcal{H}\begin{bmatrix} \psi(r;\epsilon,|\kappa|) \\ \psi(r;\epsilon,-|\kappa|) \end{bmatrix} = A^2 \begin{bmatrix} \psi(r;\epsilon,|\kappa|) \\ \psi(r;\epsilon,-|\kappa|) \end{bmatrix} = e^2 \begin{bmatrix} \psi(r;\epsilon,|\kappa|) \\ \psi(r;\epsilon,-|\kappa|) \end{bmatrix}, \qquad (14.60)$$

i.e., by solving the equation

$$A^2\psi(r;\epsilon,\kappa) = e^2\psi(r;\epsilon,\kappa) \qquad (14.61)$$

for $+\kappa$ and $-\kappa$ separately. In fact, it suffices to solve the equations for one sign. The solution corresponding to the other sign may then be obtained by Eq. (14.34).

Instead of solving Eq. (14.61) for a given value of κ, we may, of course, focus on the radial equations for the same κ-value, and this is what we shall do in the following chapter. In either case it is advantageous to exploit the way supersymmetry reflects itself in an extended function space.

Notes on Chapter 14

The Johnson–Lippmann operator was announced as a conserved quantity of the Dirac equation by Johnson and Lippmann [JL50]. The realization that it reduces to the Laplace vector in the non-relativistic limit is due to Biedenharn [Bie83]. That the Johnson–Lippmann operator generates a supersymmetry group was realized by Nieto, Kostelecky and Traux [NKT85]. The parallelism with the non-relativistic case was pointed out by Dahl and Jørgensen [DJ95].

Chapter 15
Extending the Solution Space

We may obtain a representation of the Dirac wave functions by the transformation $\psi \to \bar{\psi} = \gamma_5 \psi$. It turns out that the supersymmetry of the Dirac equation manifests itself in the new representation as a relation of the radial wave functions. We can solve this relation for the radial wave functions, and then transform back to the original representation to find the Dirac wave functions. A version of this procedure has been known for some time [Bie62] — it was found by manipulation of the solutions of Kramer's equation (see Chapter 17 below). Now that we recognize the role played by supersymmetry we can obtain the Dirac wave functions directly, without going through the detour of Kramer's equation.

In the same way that the Dirac equation leads to the Hamiltonian H of Eq. (13.13), the unphysical equation Eq. (13.5) leads to the Hamiltonian

$$\bar{H} = \boldsymbol{\alpha} \cdot \mathbf{p} - \beta m - \frac{Z\alpha}{r}, \tag{15.1}$$

which results from the ordinary Dirac Hamiltonian (13.15) by changing the sign in front of m. It is referred to as the negative-mass Hamiltonian.

It is readily seen that

$$\bar{H} = \gamma_5 H \gamma_5. \tag{15.2}$$

If, therefore, ψ is a solution of the positive-mass Eq. (13.14), then

$$\bar{\psi} = \gamma_5 \psi \tag{15.3}$$

is a solution of the negative-mass equation

$$\bar{H}\bar{\psi} = E\bar{\psi}, \tag{15.4}$$

with the same energy E.

γ_5 commutes with $\mathbf{J}$, but it anticommutes with K. This implies that if ψ is characterized by the quantum numbers j, m and κ then $\bar{\psi}$ has the quantum numbers

j, m and $-\kappa$. Thus, the sign of κ is reversed. The $\bar{\psi}$-functions differing merely in the sign of κ are transformed into each other by the operator $\bar{A} = \gamma_5 A \gamma_5$ in a similar way as in Eq. (14.34).

It follows that the stationary bound states of the negative-mass Hamiltonian define a supersymmetric structure that is, in all details, the same as the supersymmetric structure defined by the positive-mass Hamiltonian, except that the roles of κ and $-\kappa$ are reversed.

If we now group the positive-mass states with a given κ-value together with the negative-mass states with the *same* κ-value, then we obviously encounter a double stack of exactly the same appearance as before. The question is then: Is it possible to introduce a parity operator and generators such that this double stack can be properly described by supersymmetry? It turns out that this is indeed the case.

When we group the positive- and negative-mass solutions together then the states with the *same* κ-value form a new supersymmetric structure. The appearance of this structure is a direct consequence of the supersymmetric structure that combines states with *opposite* values of κ in the original Dirac–Kepler problem. The latter structure was based on the presence of the Johnson–Lippmann operator. We expect, accordingly, that this operator also plays a central role for the description of the new structure.

There is a rather direct way of carrying a positive-mass state with a given κ-value into a negative-mass state with the same κ-value, namely to apply the operators A and γ_5 in succession. Diagonalizing the operator $\gamma_5 A$ will therefore separate the double stack into two columns with different eigenvalues of $\gamma_5 A$. These columns are not affected if the operator is multiplied by an arbitrary constant and a diagonal operator added so as to obtain an operator whose eigenvalues merely differ in sign. Such an operator defines a grading of the function space and, hence, a center for the description of the supersymmetry.

Using the explicit form of the operator A, Eq. (14.18), we find its action on Dirac spinors:

$$
A \begin{pmatrix} R_A(r) \mathcal{Y}_A \\ i R_B(r) \mathcal{Y}_B \end{pmatrix} = \begin{pmatrix} \left(R_A(r) - \dfrac{\kappa}{Z\alpha}(1 - \epsilon) R_B(r) \right) \mathcal{Y}_B \\ i \left(R_B(r) - \dfrac{\kappa}{Z\alpha}(1 + \epsilon) R_A(r) \right) \mathcal{Y}_A \end{pmatrix}. \tag{15.5}
$$

Applying the matrix form of γ_5, Eq. (13.7), yields

$$
\gamma_5 \begin{pmatrix} R_A(r) \mathcal{Y}_A \\ i R_B(r) \mathcal{Y}_B \end{pmatrix} = \begin{pmatrix} i R_B(r) \mathcal{Y}_B \\ R_A(r) \mathcal{Y}_A \end{pmatrix}. \tag{15.6}
$$

Hence

$$-iZ\alpha\gamma_5 A \begin{pmatrix} R_A(r)\mathcal{Y}_A \\ iR_B(r)\mathcal{Y}_B \end{pmatrix} = iZ\alpha \begin{pmatrix} i\left(-R_B(r) + \dfrac{\kappa}{Z\alpha}(1-\epsilon)R_A(r)\right)\mathcal{Y}_A \\ \left(-R_A(r) + \dfrac{\kappa}{Z\alpha}(1+\epsilon)R_B(r)\right)\mathcal{Y}_B \end{pmatrix}.$$

$$(15.7)$$

Since the spinors before and after the transformation have the same value of κ, they also have the same spin and angular variable dependence. The transformation can then be expressed as a transformation of the radial functions, i.e., as a linear transformation in the space of functions $\mathcal{V}_R$ spanned by $R_A(r)$ and $R_B(r)$. That is

$$-iZ\alpha\gamma_5 A \begin{pmatrix} R_A \\ R_B \end{pmatrix} = \begin{pmatrix} \kappa(\epsilon-1) & Z\alpha \\ -Z\alpha & \kappa(\epsilon+1) \end{pmatrix} \begin{pmatrix} R_A \\ R_B \end{pmatrix}. \qquad (15.8)$$

The eigenvalues of this matrix are, by the Cayley characteristic equation, the roots of

$$\det \begin{pmatrix} \kappa(\epsilon-1)-\lambda & Z\alpha \\ -Z\alpha & \kappa(\epsilon+1)-\lambda \end{pmatrix} = 0, \qquad (15.9)$$

i.e., $\lambda = \kappa\epsilon \pm \gamma|\kappa|$. The grading may therefore be defined by

$$\Gamma = -\frac{KH}{m} - iZ\alpha\gamma_5 A, \qquad (15.10)$$

which has the eigenvalues $\pm\gamma|\kappa|$.

15.1 The Γ-Induced Radial Supersymmetry

The operator $\Gamma = -KH/m - iZ\alpha\gamma_5 A$ induces the following linear transformation in the space $\mathcal{V}_R$:

$$\Gamma(u_A, u_B) = (u_A, u_B) \begin{pmatrix} -\kappa & -Z\alpha \\ Z\alpha & \kappa \end{pmatrix}. \qquad (15.11)$$

We recognize this as the matrix in the first term of the Dirac equation for the radial functions R_A and R_B, Eq. (13.65). This matrix is diagonalized by

$$D = \begin{pmatrix} (Z\alpha)c_1 & (Z\alpha)c_2 \\ -(\gamma|\kappa|+\kappa)c_1 & (\gamma|\kappa|-\kappa)c_2 \end{pmatrix}, \qquad (15.12)$$

with c_1, c_2 arbitrary constants. Then

$$D^{-1} = \frac{1}{\det} \begin{pmatrix} (\gamma|\kappa|-\kappa)c_2 & -(Z\alpha)c_2 \\ (\gamma|\kappa|-\kappa)c_1 & (Z\alpha)c_1 \end{pmatrix}, \qquad (15.13)$$

with

$$\det = 2c_1 c_2 (Z\alpha)\gamma|\kappa|, \qquad (15.14)$$

and

$$D^{-1}\Gamma D = \begin{pmatrix} \gamma|\kappa| & 0 \\ 0 & -\gamma|\kappa| \end{pmatrix}. \qquad (15.15)$$

The eigenvalues of Γ are thus

$$\gamma_1 = \gamma|\kappa|, \quad \gamma_2 = -\gamma|\kappa|, \qquad (15.16)$$

where γ is as in Eq. (14.50). The grading operator is $P_\kappa = \Gamma/\gamma|\kappa|$. The new basis vectors are

$$(w_1, w_2) = D(u_A, u_B) = (u_A, u_B) \begin{pmatrix} (Z\alpha)c_1 & (Z\alpha)c_2 \\ -(\gamma|\kappa| + \kappa)c_1 & (\gamma|\kappa| - \kappa)c_2 \end{pmatrix}. \qquad (15.17)$$

The vector $(w_1, 0)$ is an eigenvector of Γ, referred to the new basis, corresponding to the eigenvalue $\gamma|\kappa|$, the vector $(0, w_2)$ is an eigenvector of Γ corresponding to the eigenvalue $-\gamma|\kappa|$. The inverse of Eq. (15.17) is

$$(u_A, u_B) = D^{-1}(w_1, w_2) = \frac{1}{\det}(w_1, w_2) \begin{pmatrix} (\gamma|\kappa| - \kappa)c_2 & -(Z\alpha)c_2 \\ (\gamma|\kappa| + \kappa)c_1 & (Z\alpha)c_2 \end{pmatrix}. \qquad (15.18)$$

We shall now perform a similarity transformation to the basis (15.17) in the Eqs. (13.65) for the radial Dirac functions du_A/dx, du_B/dx. We have still to calculate the matrix in the second term of (13.65):

$$\frac{1}{2\sqrt{1-\epsilon^2}} D^{-1} \begin{pmatrix} 0 & 1-\epsilon \\ 1-\epsilon & 0 \end{pmatrix} D$$

$$= \begin{pmatrix} \dfrac{-(Z\alpha)\epsilon}{2\gamma|\kappa|\sqrt{1-\epsilon^2}} & \dfrac{c_2}{c_1}\left[\dfrac{(\kappa-\gamma|\kappa|)^2}{4\gamma|\kappa|(Z\alpha)}\sqrt{\dfrac{1-\epsilon}{1+\epsilon}} - \dfrac{(Z\alpha)}{4\gamma|\kappa|}\sqrt{\dfrac{1+\epsilon}{1-\epsilon}}\right] \\ -\dfrac{c_1}{c_2}\left[\dfrac{(\kappa+\gamma|\kappa|)^2}{4\gamma|\kappa|(Z\alpha)}\sqrt{\dfrac{1-\epsilon}{1+\epsilon}} - \dfrac{(Z\alpha)}{4\gamma|\kappa|}\sqrt{\dfrac{1+\epsilon}{1-\epsilon}}\right] & \dfrac{(Z\alpha)\epsilon}{2\gamma|\kappa|\sqrt{1-\epsilon^2}} \end{pmatrix} \qquad (15.19)$$

The result of the similarity transformation on the differential equations for $du_A/dx, du_B/dx$ is the sum of the results for the first and second terms:

$$\left(\frac{dw_1}{dx}, \frac{dw_2}{dx}\right)$$

$$= (w_1, w_2) \begin{pmatrix} \dfrac{-Z\alpha\epsilon}{2\gamma|\kappa|\sqrt{1-\epsilon^2}} + \dfrac{\gamma|\kappa|}{x} & \dfrac{c_2}{c_1}\left[\dfrac{(\kappa-\gamma|\kappa|)^2}{4\gamma|\kappa|(Z\alpha)}\sqrt{\dfrac{1-\epsilon}{1+\epsilon}} - \dfrac{(Z\alpha)}{4\gamma|\kappa|}\sqrt{\dfrac{1+\epsilon}{1-\epsilon}}\right] \\[3mm] -\dfrac{c_1}{c_2}\left[\dfrac{(\kappa+\gamma|\kappa|)^2}{4\gamma|\kappa|(Z\alpha)}\sqrt{\dfrac{1-\epsilon}{1+\epsilon}} - \dfrac{(Z\alpha)}{4\gamma|\kappa|}\sqrt{\dfrac{1+\epsilon}{1-\epsilon}}\right] & \dfrac{(Z\alpha)\epsilon}{2\gamma|\kappa|\sqrt{1-\epsilon^2}} - \dfrac{\gamma|\kappa|}{x} \end{pmatrix}$$

$$(15.20)$$

From this we see that if we put

$$\frac{c_1}{c_2} = \sqrt{\frac{(\kappa\epsilon + \gamma|\kappa|)(\kappa - \gamma|\kappa|)}{(\kappa\epsilon - \gamma|\kappa|)(\kappa + \gamma|\kappa|)}}, \tag{15.21}$$

then we obtain the more symmetrical form:

$$\left(\frac{dw_1}{dx}, \frac{dw_2}{dx}\right) = (w_1, w_2) \begin{pmatrix} -\dfrac{\epsilon n_a}{2\gamma|\kappa|} + \dfrac{\gamma|\kappa|}{x} & -\dfrac{\epsilon n_a}{2\gamma|\kappa|} \\[3mm] \dfrac{\epsilon n_a}{2\gamma|\kappa|} & \dfrac{\epsilon n_a}{2\gamma|\kappa|} - \dfrac{\gamma|\kappa|}{x} \end{pmatrix}. \tag{15.22}$$

Here we have used

$$\frac{c_1}{c_2}\left(\frac{1}{4(Z\alpha)\gamma|\kappa|}\right) \frac{[(\kappa + \gamma|\kappa|)^2(1-\epsilon) - (Z\alpha)^2(1+\epsilon)]}{\sqrt{1-\epsilon^2}}$$

$$= -\frac{c_1}{c_2}\left(\frac{1}{2(Z\alpha)\gamma|\kappa|}\right) \frac{(\kappa + \gamma|\kappa|)(\kappa\epsilon - \gamma|\kappa|)}{\sqrt{1-\epsilon^2}} = \frac{-1}{2\gamma|\kappa|}\sqrt{\frac{(\kappa\epsilon)^2 - \gamma^2|\kappa|^2}{1-\epsilon^2}}, \tag{15.23}$$

and

$$\frac{c_2}{c_1}\left(\frac{1}{4(Z\alpha)\gamma|\kappa|}\right) \frac{(\kappa - \gamma|\kappa|)^2(1-\epsilon) - (Z\alpha)^2(1+\epsilon))}{\sqrt{1-\epsilon^2}}$$

$$= -\frac{c_2}{c_1}\left(\frac{1}{2(Z\alpha)\gamma|\kappa|}\right) \frac{(\kappa - \gamma|\kappa|)(\kappa\epsilon + \gamma|\kappa|)}{\sqrt{1-\epsilon^2}} = \frac{-1}{2\gamma|\kappa|}\sqrt{\frac{(\kappa\epsilon)^2 - \gamma^2|\kappa|^2}{1-\epsilon^2}}, \tag{15.24}$$

as well as

$$\sqrt{\frac{(\kappa\epsilon)^2 - \gamma^2|\kappa|^2}{1-\epsilon^2}} = \epsilon n_a, \tag{15.25}$$

which is a consequence of Eqs. (14.33) and (13.61). The coupled differential equations are written out as

$$-\frac{dw_1}{dx} - \left(\frac{\epsilon n_a}{2\gamma|\kappa|} - \frac{\gamma|\kappa|}{x}\right)w_1 = -\frac{\epsilon n_a}{2\gamma|\kappa|}w_2,$$

$$\frac{dw_2}{dx} - \left(\frac{\epsilon n_a}{2\gamma|\kappa|} - \frac{\gamma|\kappa|}{x}\right)w_2 = -\frac{\epsilon n_a}{2\gamma|\kappa|}w_1. \tag{15.26}$$

Before proceeding to solve the differential equations we can conclude some general results concerning the normalizations of the solutions, as well as certain expectation values. We shall make use of these results in Section 16.1.

We multiply the first of the equations in (15.26) by w_2, the second by w_1, and integrate over x. This yields

$$-\int w_2 \frac{dw_1}{dx}dx - \int w_2 \left(\frac{\epsilon n_a}{2\gamma|\kappa|} - \frac{\gamma|\kappa|}{x}\right)w_1 dx = -\frac{\epsilon n_a}{2\gamma|\kappa|}\int w_2^2 dx,$$

$$\int w_1 \frac{dw_2}{dx}dx - \int w_1 \left(\frac{\epsilon n_a}{2\gamma|\kappa|} - \frac{\gamma|\kappa|}{x}\right)w_2 = -\frac{\epsilon n_a}{2\gamma|\kappa|}\int w_1^2 dx. \tag{15.27}$$

Now

$$-\int w_2 \frac{dw_1}{dx}dx = \int w_1 \frac{dw_2}{dx}dx \tag{15.28}$$

by integration-by-parts, which is allowed for functions that fall off suitably at infinity. The other integrals on the left-hand side of the equations are also equal, so those on the right-hand side must be equal as well. Thus

$$\int w_1^2 dx = \int w_2^2 dx, \tag{15.29}$$

or

$$\langle w_1|w_1\rangle = \langle w_2|w_2\rangle. \tag{15.30}$$

We see that w_1 and w_2 are normalized in the same way.

We find a second relation for the functions w_1 and w_2 by multiplying the first equation by w_1, the second by w_2, and integrating. We get

$$-\int w_1 \frac{dw_1}{dx}dx - \int w_1 \left(\frac{\epsilon n_a}{2\gamma|\kappa|} - \frac{\gamma|\kappa|}{x}\right)w_1 dx = -\frac{\epsilon n_a}{2\gamma|\kappa|}w_1 w_2 dx,$$

$$-\int w_2 \frac{dw_2}{dx}dx - \int w_2 \left(\frac{\epsilon n_a}{2\gamma|\kappa|} - \frac{\gamma|\kappa|}{x}\right)w_2 dx = -\frac{\epsilon n_a}{2\gamma|\kappa|}w_2 w_1 dx. \tag{15.31}$$

The integral of a function times its derivative vanishes for acceptable functions. Since the remaining integrals are equal, we find

$$\left\langle w_1 \left| \frac{1}{x} \right| w_1 \right\rangle = \left\langle w_2 \left| \frac{1}{x} \right| w_2 \right\rangle. \tag{15.32}$$

We now proceed to solve the coupled differential equations of (15.26) explicitly. To display the supersymmetric structure of these equations, we define the operators

$$b^- = -\frac{d}{dx} + \left(\frac{\gamma|\kappa|}{x} - \frac{en_a}{2\gamma|\kappa|} \right), \quad b^+ = \frac{d}{dx} + \left(\frac{\gamma|\kappa|}{x} - \frac{en_a}{2\gamma|\kappa|} \right). \tag{15.33}$$

It is clear that these operators are the Hermitian conjugates of each other. The equations may then be written

$$b^- w_1 = -\frac{en_a}{2\gamma|\kappa|} w_2, \quad b^+ w_2 = -\frac{en_a}{2\gamma|\kappa|} w_1. \tag{15.34}$$

The operator b^+ is the analog of A^+: it transforms an even state into an odd state. The operator b^- is the analog of A^-, it transforms an odd state into an even state. The functions w_1 and w_2 take the place of ν_1 and ν_2, respectively. The equations of (15.34) imply that

$$b^+ b^- w_1 = \left(\frac{en_a}{2\gamma|\kappa|} \right)^2 w_1, \quad b^- b^+ w_2 - \left(\frac{en_a}{2\gamma|\kappa|} \right)^2 w_2. \tag{15.35}$$

Hence the analog of Eq. (11.12) becomes

$$\begin{bmatrix} b^+ b^- & 0 \\ 0 & b^- b^+ \end{bmatrix} \begin{bmatrix} w_1 \\ w_2 \end{bmatrix} = \left(\frac{en_a}{2\gamma|\kappa|} \right)^2 \begin{bmatrix} w_1 \\ w_2 \end{bmatrix}. \tag{15.36}$$

The eigenvalue of the supersymmetric Hamiltonian is zero when $e = 0$. Otherwise, it is positive.

15.2 The Supersymmetric Ground State in the Γ Representation

We are interested in the solution of the equations in (15.35) when e equals zero. With $e = 0$, Eq. (14.33) tells us that $n_a = |\kappa|$. The corresponding reduced energy

ϵ becomes γ, as shown in Eq. (14.59). Hence $\epsilon n_a = \gamma|\kappa|$. Inserting this result in (15.26) yields

$$-\frac{dw_1}{dx} - \left(\frac{1}{2} - \frac{\gamma|\kappa|}{x}\right)w_1 = 0, \quad \frac{dw_2}{dx} - \left(\frac{1}{2} - \frac{\gamma|\kappa|}{x}\right)w_2 = 0. \qquad (15.37)$$

The solutions of these equations are

$$w_1(x) = (const)x^{\gamma|\kappa|}e^{-x/2}, \quad w_2(x) = (const)x^{-\gamma|\kappa|}e^{x/2}. \qquad (15.38)$$

Of these, $w_2(x)$ is not a physically acceptable solution, because of its behavior at $x = \infty$. Only $w_1(x)$ is an acceptable solution, the supersymmetric ground state being

$$\begin{bmatrix} w_1 \\ w_2 \end{bmatrix} = (const)\begin{bmatrix} x^{\gamma|\kappa|}e^{-x/2} \\ 0 \end{bmatrix}. \qquad (15.39)$$

$w_1(x)$ corresponds to the eigenvalue $\gamma|\kappa|$ for the operator Γ. It is equal to x times the function $R_0(x)$ of Eq. (14.57). We determine the structure of the Dirac spinors by means of Eq. (15.18), we read off

$$u_B = \frac{Z\alpha}{\kappa - \gamma|\kappa|}u_A = \frac{\kappa + \gamma|\kappa|}{Z\alpha}u_A, \qquad (15.40)$$

in agreement with Eq. (14.56).

15.3 The General Solutions in the Γ Representation

To determine w_1 and w_2 for the higher levels, we first find that

$$b^+b^- = -\frac{d^2}{dx^2} + \frac{\gamma|\kappa|(\gamma|\kappa| - 1)}{x^2} - \frac{\epsilon n_a}{x} + \left(\frac{\epsilon n_a}{2\gamma|\kappa|}\right)^2,$$

$$b^-b^+ = -\frac{d^2}{dx^2} + \frac{\gamma|\kappa|(\gamma|\kappa| + 1)}{x^2} - \frac{\epsilon n_a}{x} + \left(\frac{\epsilon n_a}{2\gamma|\kappa|}\right)^2. \qquad (15.41)$$

We then insert these expressions into (15.35) and use relations (14.33) and (14.49) to get

$$(\epsilon n_a)^2 - (\epsilon n_a)^2 = -\gamma^2|\kappa|^2. \qquad (15.42)$$

This gives

$$\frac{d^2w_1}{dx^2} + \left\{-\frac{\gamma|\kappa|(\gamma|\kappa| - 1)}{x^2} + \frac{\epsilon n_a}{x} - \frac{1}{4}\right\}w_1 = 0,$$

$$\frac{d^2w_2}{dx^2} + \left\{-\frac{\gamma|\kappa|(\gamma|\kappa| + 1)}{x^2} + \frac{\epsilon n_a}{x} - \frac{1}{4}\right\}w_2 = 0. \qquad (15.43)$$

Next, we introduce new radial functions, $G_1(x)$ and $G_2(x)$, by writing

$$w_1(x) = x^{\gamma|\kappa|}e^{-x/2}G_1(x), \quad w_2(x) = x^{\gamma|\kappa|+1}e^{-x/2}G_2(x). \qquad (15.44)$$

These functions satisfy the equations

$$x\frac{d^2G_1}{dx^2} + (2\gamma|\kappa| - x)\frac{dG_1}{dx} + (\epsilon n_a - \gamma|\kappa|)G_1 = 0,$$

$$x\frac{d^2G_2}{dx^2} + (2\gamma|\kappa| + 2 - x)\frac{dG_2}{dx} + (\epsilon n_a - \gamma|\kappa| - 1)G_2 = 0. \qquad (15.45)$$

The regular solutions of these equations are confluent hypergeometric functions. Thus, our final result is

$$w_1(x) = N_1 x^{\gamma|\kappa|}e^{-x/2}{}_1F_1(-n_r, 2\gamma|\kappa|; x),$$

$$w_2(x) = N_2 x^{\gamma|\kappa|+1}e^{-x/2}{}_1F_1(-n_r + 1, 2\gamma|\kappa| + 2; x), \qquad (15.46)$$

with n_r the radial quantum number

$$n_r = \epsilon n_a - \gamma|\kappa|. \qquad (15.47)$$

For $w_1(x)$ to be a normalizable function, n_r must be a positive number or zero. For $w_2(x)$, we get the same condition, except that $n_r = 0$ is excluded. The number n_r counts the number of nodes of the radial function [Ros61].

We have

$$n_r + \gamma|\kappa| = \epsilon n_a = \frac{Z\alpha\epsilon}{\sqrt{1-\epsilon^2}}, \qquad (15.48)$$

so that

$$\frac{\epsilon^2}{1-\epsilon^2} = \frac{(n_r + \gamma|\kappa|)^2}{(Z\alpha)^2} \qquad (15.49)$$

and the formula for the energy levels is

$$\epsilon = \left(1 + \frac{(Z\alpha)^2}{(n_r + \gamma|\kappa|)^2}\right)^{-\frac{1}{2}}. \qquad (15.50)$$

In terms of the eigenvalues of the Hamiltonian

$$E = m\left(1 + \frac{(Z\alpha)^2}{(n_r + \gamma|\kappa|)^2}\right)^{-\frac{1}{2}}, \qquad (15.51)$$

which is the same as the Sommerfeld formula (12.27), with the identification of $|\kappa|$ with n_φ. This identification follows from a comparison of (12.8) and (12.13) with (14.50).

For $n_r = 0$, we have reproduced the result (15.39). For the other solutions, we must determine the relative values of N_1 and N_2. This is done in Appendix D. The result is

$$\frac{N_1}{N_2} = -\frac{2\gamma|\kappa|(2\gamma|\kappa| + 1)}{en_a}. \qquad (15.52)$$

Our final task is to express the original radial functions $u_A(x)$ and $u_B(x)$ in terms of $w_1(x)$ and $w_2(x)$. Let us put

$$(c_1, c_2) = \left(\frac{1}{\sqrt{(\kappa + \gamma|\kappa|)(\kappa\epsilon - \gamma|\kappa|)}}, \frac{1}{\sqrt{(\kappa - \gamma|\kappa|)(\kappa\epsilon + \gamma|\kappa|)}} \right), \qquad (15.53)$$

in accordance with Eq. (15.21). We get then, by Eq. (15.18),

$$u_A(x) = \frac{Sign(\kappa)}{2\gamma|\kappa|} \left(-\sqrt{(\kappa - \gamma|\kappa|)(\kappa\epsilon - \gamma|\kappa|)} w_1(x) \right.$$
$$\left. + \sqrt{(\kappa + \gamma|\kappa|)(\kappa\epsilon + \gamma|\kappa|)} w_2(x) \right),$$
$$u_B(x) = \frac{1}{2\gamma|\kappa|} \left(-\sqrt{(\kappa + \gamma|\kappa|)(\kappa\epsilon - \gamma|\kappa|)} w_1(x) \right. \qquad (15.54)$$
$$\left. + \sqrt{(\kappa - \gamma|\kappa|)(\kappa\epsilon + \gamma|\kappa|)} w_2(x) \right),$$

where $Sign(\kappa)$ is an overall sign factor commented on in Chapter 18. Thus, we arrive at the desired expressions:

$$u_A(x) = \mathcal{N} Sign(\kappa) x^{\gamma|\kappa|} e^{-x/2} \left[\sqrt{(\kappa - \gamma|\kappa|)(\kappa\epsilon - \gamma|\kappa|)}\,_1 \right.$$
$$\times\ F_1(-n_r, 2\gamma|\kappa|; x) + \frac{en_a x}{2\gamma|\kappa|(2\gamma|\kappa|+1)}$$
$$\left. \times\ \sqrt{(\kappa + \gamma|\kappa|)(\kappa\epsilon + \gamma|\kappa|)}\,_1 F_1(-n_r + 1, 2\gamma|\kappa| + 2; x) \right],$$
$$u_B(x) = \mathcal{N} x^{\gamma|\kappa|} e^{-x/2} \left[\sqrt{(\kappa + \gamma|\kappa|)(\kappa\epsilon - \gamma|\kappa|)}\,_1 F_1(-n_r, 2\gamma|\kappa|; x) \right. \qquad (15.55)$$
$$+ \frac{en_a x}{2\gamma|\kappa|(2\gamma|\kappa|+1)} \sqrt{(\kappa - \gamma|\kappa|)(\kappa\epsilon + \gamma|\kappa|)}\,_1$$
$$\left. \times\ F_1(-n_r + 1, 2\gamma|\kappa| + 2; x) \right].$$

The overall normalization constant corresponding to the normalization condition

$$\int [u_A^2(x) + u_B^2(x)] dx = 1, \qquad (15.56)$$

is calculated in Appendix D, Eq. (D.16).

Notes on Chapter 15

This type of extension has been widely discussed in the literature. Sukumar [Suk85a; Suk85b] found the supersymmetric structure of the equations in this decomposition, and utilized it to solve them. It was also used by Jarvis and Stedman [JS86], who solved the second-order Kramer's equation, rather than the first-order Dirac equation (see Chapter 17). The same equation was solved by Swainson and Drake [SD91], and by Goodman and Ignjatović [GI97], in an article that prompted an avalanche of papers tracing earlier work on this subject ([Sta98], [Wal79], [Kol66], [Won86], [WY82], [AB78], [Bie83]). This method leads to a different relative normalization of the solutions than that used here. The uncoupled differential equations had been previously obtained by Infeld [Inf41]. The uncoupled differential equations of (15.43) may be said to arise from the coupled equations of (15.26) by factorization, according to the general method of Hull and Infeld [HI51]. The advantage of the method used here, which utilizes the primary supersymmetry discussed in Chapter 14, is that it leads directly to the solutions of the Dirac equation, and makes clearer the nature of the coefficients appearing.

Chapter 16

A Different Extension of the Solution Space

Here we investigate another representation of the solutions, more closely related to the standard method of solution, which was actually derived in a somewhat *ad hoc* manner [Gor28]. In this representation the solutions are decomposed into *normal modes*, and can accordingly be decoupled. The presentation here, which again follows that of Dahl and Jørgensen [DJ95], profits from utilizing the supersymmetry, which is recognized already in its primary form, from the beginning.

Having extended our analysis to include the stationary bound states of the Hamiltonian $\bar{H}$, we shall now make a further, less obvious, extension. This is to consider also the stationary bound states of the Hamiltonian

$$\tilde{H} = (\mathbf{\Sigma}\cdot\hat{\mathbf{r}})H(\mathbf{\Sigma}\cdot\hat{\mathbf{r}}) = H + \frac{2}{r}\beta\gamma_5 K(\mathbf{\Sigma}\cdot\hat{\mathbf{r}}). \tag{16.1}$$

These bound states are obviously described by the functions

$$\tilde{\psi} = (\mathbf{\Sigma}\cdot\hat{\mathbf{r}})\psi, \tag{16.2}$$

where ψ describes a stationary bound state of the original Hamiltonian. Like $\beta\gamma_5$, the operator $\mathbf{\Sigma}\cdot\hat{\mathbf{r}}$ commutes with $\mathbf{J}$ and anticommutes with K. The function $\tilde{\psi}$ bears, therefore, a similar relation to ψ as does the function $\bar{\psi}$. Hence, the stationary bound states of the Hamiltonian $\tilde{H}$ also define a supersymmetric structure that is, in all details, the same as the supersymmetric structure defined by the positive-mass Hamiltonian, except that the roles of κ and $-\kappa$ are reversed.

In a similar way to what we did for the negative-mass states, we may now group the positive-mass states with a given κ-value together with the new $\tilde{\psi}$ states with the *same* κ-value and obtain a double stack of exactly the same appearance as previously. Again, we may ask if it is possible to introduce a parity operator and generators such that this double stack can be properly described by supersymmetry. We now explain how this comes about.

The function $\tilde{\psi}$, which corresponds to ψ but has opposite κ-value, is $\tilde{\psi} = (\mathbf{\Sigma \cdot \hat{r}})\psi$. The function $\bar{\psi}$ with the same κ-value as ψ is $\bar{\psi} = \mathbf{\Sigma \cdot \hat{r}} A \psi$, since the functions $\tilde{\psi}$ that differ from each other in the sign of κ are related by the operator $\tilde{A} = (\mathbf{\Sigma \cdot \hat{r}}) A (\mathbf{\Sigma \cdot \hat{r}})$. We recall Eq. (15.5):

$$A \begin{pmatrix} R_A(r)\mathcal{Y}_A \\ iR_B(r)\mathcal{Y}_B \end{pmatrix} = \begin{pmatrix} \left(R_A(r) - \dfrac{\kappa}{Z\alpha}(1+\epsilon)R_B(r) \right) \mathcal{Y}_B \\ i \left(R_B(r) - \dfrac{\kappa}{Z\alpha}(1-\epsilon)R_A(r) \right) \mathcal{Y}_A \end{pmatrix}, \qquad (16.3)$$

and Eq. (14.44):

$$\mathbf{\Sigma \cdot \hat{r}} \begin{pmatrix} R_A(r)\mathcal{Y}_A \\ iR_B(r)\mathcal{Y}_B \end{pmatrix} = - \begin{pmatrix} R_A(r)\mathcal{Y}_B \\ iR_B(r)\mathcal{Y}_A \end{pmatrix}. \qquad (16.4)$$

We see that both these operators interchange $\mathcal{Y}_A$ and $\mathcal{Y}_B$. By applying the two operators in succession we preserve the dependence on the angular variables and spin. Since both $\mathbf{\Sigma \cdot \hat{r}}$ and A commute with $\mathbf{J}$ but anticommute with K, when we apply $(\mathbf{\Sigma \cdot \hat{r}})A$ the eigenvalues of H and $\mathbf{J}$ are preserved, and the eigenvalue of K is reversed in sign. Applying $(\mathbf{\Sigma \cdot \hat{r}})A$ to ψ, we only encounter linear combinations of the original radial functions $R_A(r)$ and $R_B(r)$, i.e., functions belonging to the function space $\mathcal{V}_R$ spanned by $R_A(r), R_B(r)$. Thus the functions $\bar{\psi}(r; \epsilon, \kappa), \bar{\psi}(r; \epsilon, -\kappa)$ and $\tilde{\psi}(r; \epsilon, \kappa), \tilde{\psi}(r; \epsilon, -\kappa)$ belong, together with $\psi(r; \epsilon, \kappa)$ and $\psi(r; \epsilon, -\kappa)$, to the space $\mathcal{V}_R$. The supersymmetry which is expressed by these operators may again be described as a symmetry in the space $\mathcal{V}_R$.

Explicitly,

$$(\mathbf{\Sigma \cdot \hat{r}})A \begin{pmatrix} R_A(r)\mathcal{Y}_A \\ iR_B(r)\mathcal{Y}_B \end{pmatrix} = \begin{pmatrix} \left(-R_A(r) + \dfrac{\kappa}{Z\alpha}(1+\epsilon)R_B(r) \right) \mathcal{Y}_A \\ i \left(-R_B(r) + \dfrac{\kappa}{Z\alpha}(1-\epsilon)R_A(r) \right) \mathcal{Y}_B \end{pmatrix}. \qquad (16.5)$$

The representation of $(\mathbf{\Sigma \cdot \hat{r}})A$ in the space of functions $\mathcal{V}_R$ is

$$(\mathbf{\Sigma \cdot \hat{r}})A = \begin{pmatrix} -1 & \dfrac{\kappa}{Z\alpha}(1+\epsilon) \\ \dfrac{\kappa}{Z\alpha}(1-\epsilon) & -1 \end{pmatrix}. \qquad (16.6)$$

The characteristic equation of this matrix is

$$\det \begin{pmatrix} -1-\lambda & \dfrac{\kappa}{Z\alpha}(1+\epsilon) \\ \dfrac{\kappa}{Z\alpha}(1-\epsilon) & -1-\lambda \end{pmatrix} = 0. \qquad (16.7)$$

Hence its eigenvalues are

$$\lambda = -1 \pm \frac{\kappa}{n_a}. \tag{16.8}$$

Thus the operator

$$\Delta = 1 + (\mathbf{\Sigma \cdot \hat{r}})A \tag{16.9}$$

has eigenvalues $\pm \kappa / n_a$, and defines a parity operator

$$\mathcal{P}_{\kappa / n_a} = \frac{n_a \Delta}{\kappa}, \tag{16.10}$$

which determines a grading of the supersymmetry.

16.1 The Δ-Induced Radial Supersymmetry

The operator $\Delta = 1 + (\mathbf{\Sigma \cdot \hat{r}})A$ induces the following linear transformation in the space $\mathcal{V}_R$:

$$\Delta(u_A, u_B) = \frac{\kappa}{Z\alpha}(u_A, u_B) \begin{pmatrix} 0 & 1 - \epsilon \\ 1 + \epsilon & 0 \end{pmatrix}. \tag{16.11}$$

We recognize the second matrix in the Dirac radial equations, Eq. (13.65).
    This matrix is diagonalized by

$$D = \begin{pmatrix} \dfrac{d_1}{\sqrt{1 + \epsilon}} & -\dfrac{d_2}{\sqrt{1 + \epsilon}} \\ \dfrac{d_1}{\sqrt{1 - \epsilon}} & \dfrac{d_2}{\sqrt{1 - \epsilon}} \end{pmatrix}, \tag{16.12}$$

with d_1, d_2 arbitrary constants. Then

$$D^{-1} = \frac{1}{det} \begin{pmatrix} \dfrac{d_2}{\sqrt{1 - \epsilon}} & \dfrac{d_2}{\sqrt{1 + \epsilon}} \\ -\dfrac{d_1}{\sqrt{1 - \epsilon}} & \dfrac{d_1}{\sqrt{1 + \epsilon}} \end{pmatrix}, \tag{16.13}$$

with

$$det = \frac{2d_1 d_2}{\sqrt{1 - \epsilon^2}}, \tag{16.14}$$

and

$$D^{-1}\Delta D = \begin{pmatrix} \kappa / n_a & 0 \\ 0 & -\kappa / n_a \end{pmatrix}. \tag{16.15}$$

So the eigenvalues of Δ are

$$\delta_1 = \frac{\kappa}{n_a}, \quad \delta_2 = -\frac{\kappa}{n_a}. \tag{16.16}$$

The new basis vectors are

$$(u_1, u_2) = D(u_A, u_B) = (u_A, u_B) \begin{pmatrix} \dfrac{d_1}{\sqrt{1+\epsilon}} & -\dfrac{d_2}{\sqrt{1+\epsilon}} \\ \dfrac{d_1}{\sqrt{1-\epsilon}} & \dfrac{d_2}{\sqrt{1-\epsilon}} \end{pmatrix}. \tag{16.17}$$

The vector $(u_1, 0)$ is an eigenvector of Δ, referred to the new basis, corresponding to the eigenvalue κ/n_a. The vector $(0, u_2)$ is an eigenvector of Δ corresponding to the eigenvalue $-\kappa/n_a$. The inverse to Eq. (16.17) is

$$(u_A, u_B) = D^{-1}(u_1, u_2) = \frac{1}{\det} \begin{pmatrix} \dfrac{d_2}{\sqrt{1-\epsilon}} & \dfrac{d_2}{\sqrt{1+\epsilon}} \\ -\dfrac{d_1}{\sqrt{1-\epsilon}} & \dfrac{d_1}{\sqrt{1+\epsilon}} \end{pmatrix}. \tag{16.18}$$

The functions u_1 and u_2 are also the "normal modes" that decouple the differential equations that in turn ensure that u_A; u_B are solutions of the radial Dirac equations. Indeed, these differential equations become, after a similarity transformation is performed,

$$\left(\frac{du_1}{dx}, \frac{du_2}{dx} \right) = (u_1, u_2) \begin{pmatrix} \dfrac{1}{2} - \dfrac{\epsilon n_a}{x} & -\dfrac{d_2}{d_1} \dfrac{(n_a + \kappa)}{x} \\ \dfrac{d_1}{d_2} \dfrac{(n_a + \kappa)}{x} & \dfrac{1}{2} - \dfrac{a}{x} \end{pmatrix}. \tag{16.19}$$

Choosing

$$\frac{d_1}{d_2} = \sqrt{\frac{n_a - \kappa}{n_a + \kappa}} \tag{16.20}$$

leads to the more symmetrical equations

$$\left(\frac{du_1}{dx}, \frac{du_2}{dx} \right) = (u_1, u_2) \begin{pmatrix} \dfrac{1}{2} - \dfrac{\epsilon n_a}{x} & -\dfrac{\epsilon n_a}{x} \\ \dfrac{\epsilon n_a}{x} & -\dfrac{1}{2} + \dfrac{\epsilon n_a}{x} \end{pmatrix}, \tag{16.21}$$

where we have used

$$e^2 n_a^2 = n_a^2 - \kappa^2. \tag{16.22}$$

These equations can be written out as

$$-\frac{du_1}{dx} + \left(\frac{1}{2} - \frac{\epsilon n_a}{x}\right) u_1 = -\frac{\epsilon n_a}{x} u_2,$$

$$\frac{du_2}{dx} + \left(\frac{1}{2} - \frac{\epsilon n_a}{x}\right) u_2 = -\frac{\epsilon n_a}{x} u_1. \tag{16.23}$$

Let us multiply the first of these equations from the left by u_2, the second by u_1, and integrate over x. The left-hand sides of the equations will then be the same for acceptable radial functions, since $-(d/dx)$ is the Hermitian adjoint of (d/dx). The right-hand sides must therefore also be the same. Hence, we get

$$\left\langle u_1 \left| \frac{1}{x} \right| u_1 \right\rangle = \left\langle u_2 \left| \frac{1}{x} \right| u_2 \right\rangle. \tag{16.24}$$

Next, let us multiply the first of the equations of (16.23) from the left by u_1, the second by u_2, and integrate over x. We then compare the resulting equations, noting that the integral of a function times its derivative vanishes for an acceptable function. Thus, we get

$$\langle u_1 | u_1 \rangle = \langle u_2 | u_2 \rangle. \tag{16.25}$$

Now replace the equations of (16.23) with the equations obtained from them by multiplying them from the left by x. This gives

$$-x\frac{du_1}{dx} + \left(\frac{x}{2} - \epsilon n_a\right) u_1 = -\epsilon n_a u_2,$$

$$x\frac{du_2}{dx} + \left(\frac{x}{2} - \epsilon n_a\right) u_2 = -\epsilon n_a u_1. \tag{16.26}$$

Let us multiply the first of these equations from the left by u_2, the second by u_1, and integrate over x. A comparison of the resulting equations gives

$$\left\langle u_2 \left| x\frac{d}{dx} \right| u_1 \right\rangle = \left\langle u_1 \left| -x\frac{d}{dx} \right| u_2 \right\rangle. \tag{16.27}$$

The Hermitian adjoint of $x(d/dx)$ is $-(d/dx)x = -x(d/dx) - 1$. Hence, we conclude that u_1 and u_2 are orthogonal:

$$\langle u_1 | u_2 \rangle = 0. \tag{16.28}$$

This relation justifies the designation *normal modes* for the functions $u_1(x)$ and $u_2(x)$.

The equations in (16.26) display the supersymmetric structure of our problem. For if we introduce the operators a^- and a^+ by the definitions

$$a^- = -x\frac{d}{dx} + \left(\frac{x}{2} - \epsilon n_a\right),$$

$$a^+ = x\frac{d}{dx} + \left(\frac{x}{2} - \epsilon n_a\right),$$

(16.29)

then Eq. (16.27) allows us to consider these operators to be the Hermitian adjoints of each other. Hence, they may take the place of the operators A^- and A^+ in Section 11.1, while u_1 and u_2 take the place of v_1 and v_2, respectively. A supersymmetric Hamiltonian may thus be defined as in Eq. (11.5). The equations of (16.23) may thus be written as

$$a^- u_1 = -\epsilon n_a u_2,$$

$$a^+ u_2 = -\epsilon n_a u_1.$$

(16.30)

They show that

$$a^+ a^- u_1 = (\epsilon n_a)^2 u_1,$$

$$a^- a^+ u_2 = (\epsilon n_a)^2 u_2.$$

(16.31)

Hence, the analog of Eq. (16.30) becomes

$$\begin{bmatrix} a^+ a^- & 0 \\ 0 & a^- a^+ \end{bmatrix} \begin{bmatrix} u_1 \\ u_2 \end{bmatrix} = (\epsilon n_a)^2 \begin{bmatrix} u_1 \\ u_2 \end{bmatrix}.$$

(16.32)

Thus, the quantity $(\epsilon n_a)^2$ replaces the quantity $\mathcal{E}$ as the eigenvalue of the supersymmetric Hamiltonian. It is positive or zero, as it should be, and it is a well-defined function of the physical energy. It becomes zero when $e = 0$.

16.2 The Supersymmetric Ground State in the Δ Representation

The supersymmetric ground state is the solution of the equations in (16.30) when e equals zero. With $e = 0$, Eq. (14.33) tells us that $n_a = |\kappa|$. The corresponding reduced energy ϵ becomes γ, as shown in Eq. (14.59). Hence, $\epsilon n_a = \gamma|\kappa|$. Inserting this result in the expressions for a^- and a^+ gives the equations

$$-x\frac{du_1}{dx} + \left(\frac{1}{2}x - \gamma|\kappa|\right) u_1 = 0,$$

$$x\frac{d_2}{dx} + \left(\frac{1}{2}x - \gamma|\kappa|\right) u_2 = 0.$$

(16.33)

The solutions of these equations are

$$u_1(x) = (const)x^{-\gamma|\kappa|}e^{x/2},$$
$$u_2(x) = (const)x^{\gamma|\kappa|}e^{-x/2}. \tag{16.34}$$

Of these, $u_1(x)$ is not a physically acceptable solution. The supersymmetric ground state is accordingly

$$\begin{bmatrix} u_1 \\ u_2 \end{bmatrix} = (const) \begin{bmatrix} 0 \\ x^{\gamma|\kappa|}e^{-x/2} \end{bmatrix}. \tag{16.35}$$

The function $u_2(x)$ equals x times the $R_0(x)$ of Eq. (14.57), as we would expect. From Eq. (16.18) we find

$$\frac{u_B}{u_A} = -\sqrt{\frac{1-\epsilon}{1+\epsilon}} = -\sqrt{\frac{1-\gamma}{1+\gamma}}$$
$$= -\frac{\sqrt{1-\gamma^2}}{1+\gamma} = -\frac{Z\alpha}{|\kappa|(1+\gamma)}$$
$$= \frac{-|\kappa|+\gamma|\kappa|}{Z\alpha} = \frac{\kappa+\gamma|\kappa|}{Z\alpha}, \tag{16.36}$$

since κ is negative, in accordance with Eq. (14.56).

16.3 The General Solutions in the Δ Representation

To determine $u_1(x)$ and $u_2(x)$ for the higher levels, we must solve the uncoupled differential equations in (16.31). It is easily seen that

$$a^+a^- = -x^2\frac{d^2}{dx^2} - x\frac{d}{dx} + \frac{1}{4}x^2 - \left(\epsilon n_a - \frac{1}{2}\right)x + (\epsilon n_a)^2,$$
$$a^-a^+ = -x^2\frac{d^2}{dx^2} - x\frac{d}{dx} + \frac{1}{4}x^2 - \left(\epsilon n_a + \tfrac{1}{2}\right)x + (\epsilon n_a)^2. \tag{16.37}$$

Hence, the differential equations in (16.30), with the help of Eq. (15.42), become

$$x^2\frac{d^2u_1}{dx^2} + x\frac{du_1}{dx} + \left[-\frac{1}{4}x^2 + \left(\epsilon n_a - \frac{1}{2}\right)x - \gamma^2|\kappa|^2\right]u_1 = 0,$$
$$x^2\frac{d^2u_2}{dx^2} + x\frac{du_2}{dx} + \left[-\frac{1}{4}x^2 + \left(\epsilon n_a + \frac{1}{2}\right)x - \gamma^2|\kappa|^2\right]u_2 = 0. \tag{16.38}$$

The substitutions

$$u_1(x) = x^{\gamma|\kappa|}e^{-x/2}F_1(x),$$
$$u_2(x) = x^{\gamma|\kappa|}e^{-x/2}F_2(x),$$

(16.39)

introduce new radial functions, proportional to $F_1(x)$ and $F_2(x)$, that must satisfy the equations

$$x\frac{d^2F_1}{dx^2} + (2\gamma|\kappa| + 1 - x)\frac{dF_1}{dx} + (\epsilon n_a - \gamma|\kappa| - 1)F_1 = 0,$$
$$x\frac{d^2F_2}{dx^2} + (2\gamma|\kappa| + 1 - x)\frac{dF_2}{dx} + (\epsilon n_a - \gamma|\kappa|)F_2 = 0.$$

(16.40)

The regular solutions of these equations are confluent hypergeometric functions; hence, we get

$$u_1(x) = N_1 x^{\gamma|\kappa|}e^{-x/2}\,_1F_1(-n_r + 1, 2\gamma|\kappa| + 1; x),$$
$$u_2(x) = N_2 x^{\gamma|\kappa|}e^{-x/2}\,_1F_1(-n_r, 2\gamma|\kappa| + 1; x),$$

(16.41)

where n_r is the radial quantum number:

$$n_r = \epsilon n_a - \gamma|\kappa|.$$

(16.42)

For $u_2(x)$ to be a normalizable function, n_r must be a positive integer or zero. For $u_1(x)$, we get the same condition, except that $n_r = 0$ is excluded. Thus, we obtain the energy levels described in Chapter 15.

For $n_r = 0$, we have now reproduced the result (16.35). For the other solutions, we must determine the relative values of N_1 and N_2. We already know, from Eq. (16.25), that $u_1(x)$ and $u_2(x)$ must be normalized in the same way, but we do not know the relative signs of the functions. By explicit calculation (see Appendix E) we get

$$\frac{N_1}{N_2} = \frac{n_r}{\sqrt{(n_a + \kappa)(n_a - \kappa)}}.$$

(16.43)

We have now determined the complete solutions of the supersymmetric eigenvalue problem (16.32). We now use the fact that $u_1(x)$ and $u_2(x)$ are connected to the functions $u_a(x)$ and $u_B(x)$ through the relations (16.17) and (16.20). We can therefore obtain the analytical expression for $u_A(x)$ and $u_B(x)$ from Eq. (16.17). We put

$$(d_1, d_2) = \left(1, \sqrt{\frac{n_a + \gamma|\kappa|}{n_a - \gamma|\kappa|}}\right),$$

(16.44)

in accordance with Eq. (16.20), and get

$$
u_A(x) = \frac{1}{2}\sqrt{1+\epsilon}\left(u_1(x) - \sqrt{\frac{n_a - \gamma|\kappa|}{n_a + \gamma|\kappa|}}\, u_2(x)\right),
$$
$$
u_B(x) = \frac{1}{2}\sqrt{1-\epsilon}\left(u_1(x) + \sqrt{\frac{n_a - \gamma|\kappa|}{n_a + \gamma|\kappa|}}\, u_2(x)\right).
$$
(16.45)

The final expressions become

$$
u_A(x) = \frac{\mathcal{N}'}{2}\sqrt{1+\epsilon}\,x^{\gamma|\kappa|}e^{-x/2}[n_r\,{}_1F_1(-n_r + 1, 2\gamma|\kappa| + 1; x)
$$
$$
- (n_a - \gamma|\kappa|)\,{}_1F_1(-n_r, 2\gamma|\kappa| + 1; x)],
$$
$$
u_B(x) = \frac{\mathcal{N}'}{2}\sqrt{1-\epsilon}\,x^{\gamma|\kappa|}e^{-x/2}[n_r\,{}_1F_1(-n_r + 1, 2\gamma|\kappa| + 1; x)
$$
$$
+ (n_a - \gamma|\kappa|)\,{}_1F_1(-n_r, 2\gamma|\kappa| + 1; x)].
$$
(16.46)

The overall normalization constant is

$$
\mathcal{N}' = \frac{1}{\Gamma(2\gamma|\kappa| + 1)}\left[\frac{\Gamma(n_r + 2\gamma|\kappa| + 1)}{n_a(n_a - \kappa)n_r!}\right]^{1/2},
$$
(16.47)

see Eq. (E.12).

Notes on Chapter 16

The standard method of solving the radial equations (13.64) is by introducing a normal-mode representation. This method was introduced by Gordon [Gor28]. It is also used by Bethe and Salpeter [BS57]. We now understand that in these works one is actually diagonalizing the operator Δ, without paying attention to the underlying supersymmetry. With proper attention paid to this symmetry the procedure is simplified and the algebra becomes more transparent. The decoupled equations (16.38) may be said to arise from the coupled equations (16.23) by factorization.

Chapter 17

The Relation of the Solutions to Kramer's Equation

The solutions of the Dirac equation have been widely investigated in the framework of a second-order differential equation called Kramer's equation. The solutions of Kramer's equation are superpositions of the solutions of Dirac's equation and the solutions of an unphysical equation. When we have found the solutions of Kramer's equation we still have to project out the solutions of Dirac's equation [MG58]. This used to involve additional work. But in the present approach it turns out that this work is already done.

In this chapter we work out the relations between the solutions given here and those obtained by the solution of Kramer's equation. We find that, when Kramer's equation is expressed in terms of the quantity Γ, it reduces to the differential equations for w_1, w_2. When it is expressed in terms of Δ, it reduces to the equations for u_1, u_2. So we can also find the solutions to the Dirac equation, u_A, u_B, by solving Kramer's equation for w_1, w_2, or u_1, u_2, and then using the known relations between w_1, w_2 and u_A, u_B, or between u_1, u_2 and u_A, u_B. In the following sections we shall see how this works out in detail.

17.1 The Eigenvalue Problem for Traceless 2×2 Matrices

Let us consider the linear space spanned by two functions, F_1 and F_2. We introduce a linear operator Ω defined by the relation

$$\Omega(F_1, F_2) = (F_1, F_2) \begin{pmatrix} \Omega_{11} & \Omega_{12} \\ \Omega_{21} & \Omega_{22} \end{pmatrix}, \tag{17.1}$$

and assume that

$$S = c_1 F_1 + c_2 F_2 \tag{17.2}$$

is an eigenfunction of this operator with eigenvalue λ. We have then

$$\begin{pmatrix} \Omega_{11} & \Omega_{12} \\ \Omega_{21} & \Omega_{22} \end{pmatrix} \begin{pmatrix} c_1 \\ c_2 \end{pmatrix} = \lambda \begin{pmatrix} c_1 \\ c_2 \end{pmatrix}. \tag{17.3}$$

We want to determine coefficients (a_1, a_2) and (b_1, b_2), which enable us to write S as

$$S = a_1 F_1 + a_2 \Omega F_1$$
$$S = b_1 F_2 + b_2 \Omega F_2. \tag{17.4}$$

By using Eq. (17.1) we get

$$S = a_1 F_1 + a_2(\Omega_{11} F_1 + \Omega_{12} F_2) = (a_1 + a_2 \Omega_{11}) F_1 + a_2 \Omega_{12} F_2. \tag{17.5}$$

Thus, a_1 and a_2 may be determined from the equations

$$c_1 = a_1 + a_2 \Omega_{11},$$
$$c_2 = a_2 \Omega_{12}. \tag{17.6}$$

This gives

$$a_1 = c_1 - \frac{c_2 \Omega_{11}}{\Omega_{12}},$$

$$a_2 = \frac{c_2}{\Omega_{12}}, \tag{17.7}$$

and, hence,

$$S = a_2 \left[\left(\Omega_{12} \frac{c_1}{c_2} - \Omega_{11} \right) F_1 + \Omega F_1 \right]. \tag{17.8}$$

In a similar way, we get

$$S = b_1 F_2 + b_2(\Omega_{21} F_1 + \Omega_{22} F_2), \tag{17.9}$$

so that

$$c_1 = b_2 \Omega_{21},$$
$$c_2 = b_1 + b_2 \Omega_{22}, \tag{17.10}$$

or

$$b_1 = c_2 - \frac{c_1 \Omega_{22}}{\Omega_{21}},$$

$$b_2 = \frac{c_1}{\Omega_{21}}, \tag{17.11}$$

and

$$S = b_2 \left[\left(\Omega_{21} \frac{c_2}{c_1} - \Omega_{22} \right) F_2 + \Omega F_2 \right]. \tag{17.12}$$

Now, Eq. (17.3) allows us to write

$$\lambda = \Omega_{11} + \Omega_{21} \frac{c_2}{c_1} = \Omega_{22} + \Omega_{12} \frac{c_1}{c_2}. \tag{17.13}$$

We conclude, therefore, that if the matrix is traceless,

$$\Omega_{11} = -\Omega_{22}, \tag{17.14}$$

then we have the following simple result:

$$S = a_2(\lambda F_1 + \Omega F_1) \quad \text{and}$$
$$S = b_2(\lambda F_2 + \Omega F_2). \tag{17.15}$$

By taking the difference between the two expressions for S, we finally get

$$\Omega(a_2 F_1 - b_2 F_2) = -\lambda(a_2 F_1 - b_2 F_2). \tag{17.16}$$

Hence, $(a_2, -b_2)$ is an eigenvector of the matrix in Eq. (17.3) with eigenvalue $-\lambda$. That $-\lambda$ is an eigenvalue when λ is, is a consequence of the assumption that the matrix is traceless, Eq. (17.14).

17.2 Eigenfunctions of the Operators Γ and Δ

In the previous sections, we diagonalized the operators Γ and Δ in the space of functions $\mathcal{V}_R$ spanned by the radial functions $R_A(r)$ and $R_B(r)$ that belong to the Dirac spinor ψ characterized by the quantum number κ and the reduced energy ϵ. Let us now, with reference to the discussion in Section 17.1, determine the eigenfunctions of Γ in the space spanned by ψ and the spinor $\bar{\psi}$ defined by Eq. (15.3). Let us, similarly, determine the eigenfunctions of Δ in the space spanned by ψ and the spinor $\tilde{\psi}$ defined by Eq. (16.2). With reference to Eqs. (15.16) and (16.16), we denote the eigenfunctions in question by $\psi_{\gamma_1}, \psi_{\gamma_2}, \psi_{\delta_1}$, and ψ_{δ_2}. To determine them, we need some special relations that hold for solutions of eigenvalue problems of the kind discussed in the previous sections, the salient feature of these problems being that the two-dimensional matrix to be diagonalized has zero trace.

The radial eigenfunctions of the operator Γ may be read off from the equations of (15.18):

$$\begin{pmatrix} w_1 \\ w_2 \end{pmatrix} = D \begin{pmatrix} u_A \\ u_B \end{pmatrix} = \begin{pmatrix} c_1 Z\alpha & -c_1(\kappa + \gamma|\kappa|) \\ c_2(Z\alpha) & c_2(-\kappa + \gamma|\kappa|) \end{pmatrix} \begin{pmatrix} u_A \\ u_B \end{pmatrix}. \qquad (17.17)$$

Each of them parallels the function S in Eq. (17.2). That function may also be expressed in the form (17.4), with factors a_2 and b_2 as described in the comments to Eq. (17.16); after identifying the vectors F_1 and F_2 of the previous section with the functions u_A and u_B we find that we may write the radial function corresponding to the eigenvalue $\gamma_2 = -\gamma|\kappa|$ as

$$w_2 = a_2 u_a - b_2 u_B = c_2((Z\alpha)u_A + (-\kappa + \gamma|\kappa|)u_B). \qquad (17.18)$$

Therefore

$$a_2 = c_2(Z\alpha), \quad b_2 = c_2(\kappa - \gamma|\kappa|). \qquad (17.19)$$

Hence, we get that the eigenfunction corresponding to the eigenvalue $\gamma_1 = \gamma|\kappa|$ is, according to Eq. (17.15):

$$w_1 = a_2(\lambda u_A + \Gamma u_A) = c_2(Z\alpha)(\gamma|\kappa|u_A + \Gamma u_A) \quad \text{or}$$
$$w_1 = b_2(\lambda u_B + \Gamma u_B) = c_2(\kappa - \gamma|\kappa|)(\gamma|\kappa|u_B + \Gamma u_B). \qquad (17.20)$$

Accordingly, we get that

$$\gamma|\kappa|u_A(x) + \Gamma u_A(x) = \frac{1}{Z\alpha c_2}\, w_1(x),$$

$$\gamma|\kappa|u_B(x) + \Gamma u_B(x) = \frac{1}{(\kappa - \gamma|\kappa|)c_2}\, w_1(x). \qquad (17.21)$$

The spinor ψ_{γ_1} is accordingly given by

$$x\psi_{\gamma_1}(x) = N^{(1)}w_1(x) \begin{pmatrix} (\kappa - \gamma|\kappa|)\,\mathcal{Y}_A \\ iZ\alpha\,\mathcal{Y}_B \end{pmatrix}, \qquad (17.22)$$

where $N^{(1)}$ is an arbitrary constant.

In a similar way, we find that the radial function corresponding to the eigenvalue $\gamma_2 = -\gamma|\kappa|$ may be written as

$$w_2 = c_1(Z\alpha)(-\gamma|\kappa|u_A + \Gamma u_A), \quad \text{or}$$
$$w_2 = c_1(\kappa + \gamma|\kappa|)(-\gamma|\kappa|u_B + \Gamma u_B); \tag{17.23}$$

hence, we get that

$$x\psi_{\gamma_2}(x) = N^{(2)}w_2(x)\begin{pmatrix} (\kappa + \gamma|\kappa|)\,\mathcal{Y}_A \\ iZ\alpha\,\mathcal{Y}_B \end{pmatrix}. \tag{17.24}$$

To determine the spinors ψ_{δ_1} and ψ_{δ_2}, we exploit the expression (16.12), which allows us to write for the radial function corresponding to the eigenvalue $\delta_1 = \kappa/n_a$ the expressions

$$u_1 = -\frac{d_2}{\sqrt{1+\epsilon}}\left(\frac{\kappa}{n_a}u_A + \Delta u_A\right), \quad \text{or}$$
$$u_1 = -\frac{d_2}{\sqrt{1-\epsilon}}\left(\frac{\kappa}{n_a}u_B + \Delta u_B\right). \tag{17.25}$$

This gives that

$$x\psi_{\delta_1}(x) = N^{(3)}u_1(x)\begin{pmatrix} \sqrt{1+\epsilon}\,\mathcal{Y}_A \\ i\sqrt{1-\epsilon}\,\mathcal{Y}_B \end{pmatrix}. \tag{17.26}$$

Finally, we get for the radial function corresponding to the eigenvalue $\lambda_2 = -\kappa/n_a$ the expressions

$$u_2 = \frac{d_1}{\sqrt{1+\epsilon}}\left(-\frac{\kappa}{n_a}u_A + \Delta u_A\right), \quad \text{or}$$
$$u_2 = -\frac{d_1}{1-\epsilon}\left(-\frac{\kappa}{n_a}u_B + \Delta u_B\right), \tag{17.27}$$

and, hence,

$$x\psi_{\delta_2}(x) = N^{(4)}u_2(x)\begin{pmatrix} \sqrt{1+\epsilon}\,\mathcal{Y}_A \\ i\sqrt{1-\epsilon}\,\mathcal{Y}_B \end{pmatrix}. \tag{17.28}$$

The eigenspinors above are in general a superposition of states of the different supersymmetric parities. This is *not* the case for the ground state, which has no supersymmetric partner. In this case, and in this case alone,

$$x\psi_{\gamma_1}(x) = N^{(1)} w_1(x) \begin{pmatrix} (\kappa - \gamma|\kappa|) \, \mathcal{Y}_A \\ iZ\alpha \, \mathcal{Y}_B \end{pmatrix} = N^{(1')} w_1(x) \begin{pmatrix} \mathcal{Y}_A \\ i\frac{\kappa+\gamma|\kappa|}{Z\alpha} \, \mathcal{Y}_B \end{pmatrix},$$

(17.29)

where $w_1(x) = xR_0(x)$, and we have used Eq. (14.54). Similarly,

$$x\psi_{\delta_2}(x) = N^{(4)} u_2(x) \begin{pmatrix} \sqrt{1+\epsilon} \, \mathcal{Y}_A \\ i\sqrt{1-\epsilon} \, \mathcal{Y}_B \end{pmatrix} = N^{(4')} u_2(x) \begin{pmatrix} \mathcal{Y}_A \\ i\frac{\kappa+\gamma|\kappa|}{Z\alpha} \, \mathcal{Y}_B \end{pmatrix},$$

(17.30)

where $u_2(x) = xR_0(x)$, and we have used (16.36).

The eigenspinors that we have found above are, with respect to supersymmetric quantum mechanics, the parallels of the eigenfunctions of the Johnson–Lippmann operator A. Thus, A turns the supersymmetric partners $|\epsilon, j, m_j, \kappa\rangle$ and $|\epsilon, j, m_j, -\kappa>$ into each other (Eq. (14.34)). It has the eigenvalues $\pm e$ and the eigenfunctions (14.36). Similarly, the operator Γ turns the spinors ψ and $\tilde{\psi}$ into each other. It has the eigenvalues $\pm\gamma|\kappa|$ and the eigenfunctions (17.29) and (17.22). Finally, the operator Δ turns the spinors ψ and $\tilde{\psi}$ into each other. It has the eigenvalues $\pm\gamma|\kappa|$ and the eigenfunctions (17.22) and (17.24).

17.3 Kramer's Equation

We now consider Kramer's equation, which is a second-order equation satisfied by positive-mass and negative-mass spinors alike. To obtain it, we note that the equations $(H-E)\psi = 0$ (which follows from the Dirac equation) and $(\bar{H}-E)\tilde{\psi} = 0$ (which follows from Eq. (13.5)) may be replaced by the equations $O_+\psi = 0$ and $O_-\tilde{\psi} = 0$, where

$$O_+ = \beta(H - E) = m + \beta\gamma_5\boldsymbol{\Sigma}\cdot\mathbf{p} - \beta\left(E + \frac{Z\alpha}{r}\right),$$

$$O_- = \beta(\bar{H} - E) = -m + \beta\gamma_5\boldsymbol{\Sigma}\cdot\mathbf{p} - \beta\left(E + \frac{Z\alpha}{r}\right).$$

(17.31)

The operators O_+ and O_- commute. Both ψ and $\bar{\psi}$ will therefore satisfy the equation

$$O_-O_+\psi = \left\{-m^2 + \left(E + \frac{Z\alpha}{r}\right)^2 + \beta\gamma_5(\boldsymbol{\Sigma}\cdot\mathbf{p})\beta\gamma_5(\boldsymbol{\Sigma}\cdot\mathbf{p})\right.$$

$$\left. -\beta\gamma_5(\boldsymbol{\Sigma}\cdot\mathbf{p})\beta\left(E + \frac{Z\alpha}{r}\right) - \beta\left(E + \frac{Z\alpha}{r}\right)\beta\gamma_5(\boldsymbol{\Sigma}\cdot\mathbf{p})\right\}\psi = 0. \tag{17.32}$$

Now

$$\beta\gamma_5(\boldsymbol{\Sigma}\cdot\mathbf{p})\beta\gamma_5(\boldsymbol{\Sigma}\cdot\mathbf{p}) = -(\boldsymbol{\Sigma}\cdot\mathbf{p})(\boldsymbol{\Sigma}\cdot\mathbf{p}) = -\mathbf{p}^2, \tag{17.33}$$

and

$$-\beta\gamma_5(\boldsymbol{\Sigma}\cdot\mathbf{p})\beta\left(E + \frac{Z\alpha}{r}\right) - \beta\left(E + \frac{Z\alpha}{r}\right)\beta\gamma_5(\boldsymbol{\Sigma}\cdot\mathbf{p})$$

$$= \gamma_5(\boldsymbol{\Sigma}\cdot\mathbf{p})\left(E + \frac{Z\alpha}{r}\right) - \gamma_5(\boldsymbol{\Sigma}\cdot\mathbf{p})\left(E + \frac{Z\alpha}{r}\right)$$

$$= Z\alpha\gamma_5\left[\frac{1}{r}, (\boldsymbol{\Sigma}\cdot\mathbf{p})\right] = \frac{i}{r^2}Z\alpha\gamma_5(\boldsymbol{\Sigma}\cdot\hat{\mathbf{r}}). \tag{17.34}$$

Inserting these expressions in Eq. (17.32) yields

$$\left\{-m^2 - \mathbf{p}^2 + \left(E + \frac{Z\alpha}{r}\right)^2 + \frac{i}{r^2}Z\alpha\gamma_5\boldsymbol{\Sigma}\cdot\hat{\mathbf{r}}\right\}\psi = 0. \tag{17.35}$$

This is Kramer's equation. We note that O_- is not the Hermitian adjoint of O_+. The operator O_-O_+ is therefore not Hermitian. Although any solution of Dirac's equation is automatically a solution of Kramer's equation the converse is *not* true: A solution of Kramer's equation is a mixture of solutions of the Dirac equation for positive mass and the Dirac equation for negative mass. To get a solution of the Dirac equation for positive mass (the physical Dirac equation) we must, in the conventional procedure, project it out from the solution of Kramer's equation. This step, which is somewhat unwieldy in the usual formulation, is here superfluous, because the linear combination of solutions of the second-order equation, the confluent hypergeometric functions w_1 and w_2, or, alternatively, u_1 and u_2, which are also solutions of the Dirac equation, are already known.

In spherical polar coordinates

$$\mathbf{p}^2 = \frac{\mathbf{L}^2}{r^2} - \frac{1}{r^2}\frac{\partial}{\partial r} r^2 \frac{\partial}{\partial r}, \tag{17.36}$$

and noting that $L^2 = K(K + \beta)$, Eq. (13.28), the equation takes the form:

$$\left\{ -(m^2 - E^2) + \frac{1}{r^2}\frac{\partial}{\partial r} r^2 \frac{\partial}{\partial r} - \frac{K^2 - Z^2\alpha^2}{r^2} - \frac{\beta K}{r^2} + \frac{2Z\alpha E}{r} \right.$$
$$\left. + \frac{i}{r^2} Z\alpha\gamma_5\mathbf{\Sigma}\cdot\hat{\mathbf{r}} \right\}\psi = \left\{ -(m^2 - E^2) + \frac{1}{r^2}\frac{\partial}{\partial r} r^2 \frac{\partial}{\partial r} \right.$$
$$\left. - \frac{K^2 - Z^2\alpha^2}{r^2} + \frac{\Gamma}{r^2} + \frac{2Z\alpha E}{r} \right\}\psi = 0, \tag{17.37}$$

where

$$\Gamma = -\beta K + iZ\alpha\gamma_5(\mathbf{\Sigma}\cdot\hat{\mathbf{r}}). \tag{17.38}$$

The operator Γ agrees with the expression we have been using, Eq. (15.10):

$$\Gamma = -\frac{KH}{m} - iZ\alpha\gamma_5 A, \tag{17.39}$$

because

$$\Gamma = -\frac{KH}{m} - iZ\alpha\gamma_5\left(\frac{-i}{Z\alpha m}K\gamma_5(H - \beta m) - \mathbf{\Sigma}\cdot\hat{\mathbf{r}}\right)$$
$$= -\frac{KH}{m} - \gamma_5 K\gamma_5\left(\frac{H}{m} - \beta\right) + iZ\alpha\gamma_5(\mathbf{\Sigma}\cdot\hat{\mathbf{r}})$$
$$= -\frac{KH}{m} + K\left(\frac{H}{m} - \beta\right) + iZ\alpha\gamma_5(\mathbf{\Sigma}\cdot\hat{\mathbf{r}})$$
$$= -\beta K + iZ\alpha\gamma_5(\mathbf{\Sigma}\cdot\hat{\mathbf{r}}). \tag{17.40}$$

The operator βK anticommutes with $\gamma_5(\mathbf{\Sigma}\cdot\hat{\mathbf{r}})$, because

$$\{\beta K, \gamma_5(\mathbf{\Sigma}\cdot\hat{\mathbf{r}})\} = \beta K\gamma_5(\mathbf{\Sigma}\cdot\hat{\mathbf{r}}) + \gamma_5(\mathbf{\Sigma}\cdot\hat{\mathbf{r}})\beta K = \beta K\gamma_5(\mathbf{\Sigma}\cdot\hat{\mathbf{r}}) - \beta\gamma_5(\mathbf{\Sigma}\cdot\hat{\mathbf{r}})K$$
$$= \beta K\gamma_5(\mathbf{\Sigma}\cdot\hat{\mathbf{r}}) + \beta\gamma_5 K(\mathbf{\Sigma}\cdot\hat{\mathbf{r}}) = \beta K\gamma_5(\mathbf{\Sigma}\cdot\hat{\mathbf{r}}) - \beta K\gamma_5(\mathbf{\Sigma}\cdot\hat{\mathbf{r}})$$
$$= 0, \tag{17.41}$$

and hence, from Eq. (17.40),

$$\Gamma^2 = K^2 - (Z\alpha)^2. \tag{17.42}$$

Inserting Eq. (17.42) into Eq. (17.37) yields Kramer's equation in the form

$$\left\{ \frac{1}{r^2} \frac{\partial}{\partial r} r^2 \frac{\partial}{\partial r} - \frac{\Gamma(\Gamma - 1)}{r^2} + \frac{2Z\alpha E}{r} + (E^2 - m^2) \right\} \psi = 0. \tag{17.43}$$

We can rewrite the kinetic energy term with the identity

$$\frac{1}{r^2} \frac{\partial}{\partial r} r^2 \frac{\partial}{\partial r} = \frac{\partial^2}{\partial r^2} + 2r \frac{\partial}{\partial r} = \frac{1}{r} \frac{\partial^2}{\partial r^2} r. \tag{17.44}$$

Going over to the variable x, defined by Eq. (13.60), Eq. (17.37) then becomes

$$\left\{ x^{-1} \frac{\partial^2}{\partial x^2} x - x^{-2} \Gamma(\Gamma - 1) + \frac{\epsilon n_a}{x} - \frac{1}{4} \right\} \psi = 0. \tag{17.45}$$

Assume now that ψ is an eigenfuntion of Γ with eigenvalue $+\gamma|\kappa|$. We may then replace the operator Γ in Eq. (17.46) by its eigenvalue $\gamma|\kappa|$, by which the whole operator in front of $x\psi$ becomes independent of spin and angular coordinates. Multiplying on the left by x we get:

$$\left(\frac{\partial^2}{\partial x^2} + \frac{\gamma|\kappa|(\gamma|\kappa| + 1)}{x^2} + \frac{\epsilon n_a}{x} - \frac{1}{4} \right) x\psi = 0. \tag{17.46}$$

The equation may now be factorized, in accordance with Eq. (17.22), and we obtain a radial function that is seen to satisfy the upper equation in (15.43):

$$\frac{d^2 w_1}{dx^2} + \left\{ -\frac{\gamma|\kappa|(\gamma|\kappa| - 1)}{x^2} + \frac{\epsilon n_a}{x} - \frac{1}{4} \right\} w_1 = 0. \tag{17.47}$$

Similarly, the eigenvalue $-\gamma|\kappa|$ gives a radial function that satisfies the lower equation in (15.43):

$$\frac{d^2 w_2}{dx^2} + \left\{ -\frac{\gamma|\kappa|(\gamma|\kappa| + 1)}{x^2} + \frac{\epsilon n_a}{x} - \frac{1}{4} \right\} w_2 = 0. \tag{17.48}$$

Thus, solutions of Kramer's equation are solutions of the differential equations for $w_1(x)$ and $w_2(x)$. These are standard differential equations, which have explicit solutions in terms of confluent hypergeometric functions. These are connected to the radial solutions of the Dirac equation by the known matrix of Eq. (15.17).

The fact that the differential equations (15.43) may be derived from Kramer's equation seems to give the functions $w_1(x)$ and $w_2(x)$ an elevated status with respect to the Dirac equation. The functions $u_1(x)$ and $u_2(x)$ that arise from diagonalizing the operator Δ lead, however, to simpler expressions for $u_A(x)$ and $u_B(x)$ and, unlike $w_1(x)$ and $w_2(x)$, the functions $u_1(x)$ and $u_2(x)$ are mutually orthogonal. Accordingly, it is of interest to note that the second-order differential equations for $u_1(x)$ and $u_2(x)$ may also be derived from Kramer's equation. To show this, we insert the form of the Hamiltonian into expression (15.10) for Γ. We make the assumption that κ is a good quantum number and therefore replace K by κ in expression (17.38) for Γ. We may then show that Γ can be written

$$\Gamma = \frac{\partial}{\partial r}r - \frac{Z\alpha m}{\kappa}r[1 + (\mathbf{\Sigma}\cdot\hat{\mathbf{r}})A]. \qquad (17.49)$$

Indeed

$$(\mathbf{\Sigma}\cdot\hat{\mathbf{r}})A = \frac{-i}{Z\alpha m}(\mathbf{\Sigma}\cdot\hat{\mathbf{r}})K\gamma_5(H - \beta m) - 1, \qquad (17.50)$$

and therefore

$$1 + (\mathbf{\Sigma}\cdot\hat{\mathbf{r}})A = \frac{-i}{Z\alpha m}(\mathbf{\Sigma}\cdot\hat{\mathbf{r}})K\gamma_5(H - \beta m) = \frac{i}{Z\alpha m}(\mathbf{\Sigma}\cdot\hat{\mathbf{r}})\gamma_5 K(H - \beta m)$$

$$= \frac{i}{Z\alpha m}(\mathbf{\Sigma}\cdot\hat{\mathbf{r}})\gamma_5(H - \beta m)K = \frac{i\kappa}{Z\alpha m}(\mathbf{\Sigma}\cdot\hat{\mathbf{r}})\gamma_5(H - \beta m). \qquad (17.51)$$

Now, by Eq. (13.53),

$$(H - \beta m) = -i\gamma_5(\mathbf{\Sigma}\cdot\hat{\mathbf{r}})\left(\frac{\partial}{\partial r} + \frac{\beta K + 1}{r}\right) - \frac{Z\alpha}{r}$$

$$= -i\gamma_5(\mathbf{\Sigma}\cdot\hat{\mathbf{r}})\frac{1}{r}\left(\frac{\partial}{\partial r}r + \beta K\right). \qquad (17.52)$$

Then

$$(\mathbf{\Sigma}\cdot\hat{\mathbf{r}})\gamma_5(H - \beta m) = -\frac{i}{r}\left(\frac{\partial}{\partial r}r + \beta K\right) - (\mathbf{\Sigma}\cdot\hat{\mathbf{r}})\gamma_5\frac{Z\alpha}{r}, \qquad (17.53)$$

and

$$\frac{Z\alpha m}{\kappa}r[1 + (\mathbf{\Sigma}\cdot\hat{\mathbf{r}})A] = ir\left(-\frac{i}{r}\left(\frac{\partial}{\partial r}r + \beta K\right) - (\mathbf{\Sigma}\cdot\hat{\mathbf{r}})\gamma_5 Z\alpha\right)$$

$$= \frac{\partial}{\partial r}r + \beta K - iZ\alpha\gamma_5 r(\mathbf{\Sigma}\cdot\hat{\mathbf{r}}). \qquad (17.54)$$

With the definition from Eq. (17.38) this yields

$$\Gamma = \frac{\partial}{\partial r}r - \frac{Z\alpha m}{\kappa}r[1 + (\mathbf{\Sigma}\cdot\hat{\mathbf{r}})A]. \qquad (17.55)$$

Now introduce Δ from Eq. (16.9):

$$\Gamma = \frac{\partial}{\partial r}r - \frac{Z\alpha m}{\kappa}r\Delta. \qquad (17.56)$$

In terms of the variable x from (13.60),

$$\Gamma = \frac{\partial}{\partial x} x - \frac{n_a}{2\kappa} x \Delta. \qquad (17.57)$$

Insert this expression for Γ into Eq. (17.45), and, using Eq. (17.42), get

$$\left\{ x^{-1} \frac{\partial^2}{\partial x^2} x - x^2 (\kappa^2 - (Z\alpha)^2) + x^{-2} \frac{\partial}{\partial x} x - \frac{n_a}{2\kappa} x^{-1} \Delta + \frac{\epsilon n_a}{x} - \frac{1}{4} \right\} \psi = 0. \qquad (17.58)$$

Note that K commutes with $\Delta = 1 + (\Sigma \cdot \hat{r})A$, because K anticommutes with $\Sigma \cdot \hat{r}$ and A. Therefore we may choose ψ to be a simultaneous eigenvector of K and Δ. We may replace the operator Δ with one of its eigenvalues, κ/n_a or $-\kappa/n_a$, and multiply from the left with x^3, whereby the operator in front of $x\psi$ becomes independent of spin and angular coordinates:

$$\left\{ x^2 \frac{\partial^2}{\partial x^2} + x \frac{\partial}{\partial x} - \frac{1}{4} x^2 - \gamma^2 |\kappa|^2 + x \left(\epsilon n_a \mp \frac{1}{2} \right) \right\} (x\psi) = 0. \qquad (17.59)$$

Hence, the equation may be factorized, in accordance with Eqs. (17.26) and (17.28), and we obtain a radial function that satisfies, for the eigenvalue κ/n_a,

$$x^2 \frac{d^2 u_1}{dx^2} + x \frac{du_1}{dx} + \left[-\frac{1}{4} x^2 + \left(\epsilon n_a - \frac{1}{2} \right) x - \gamma^2 |\kappa|^2 \right] u_1 = 0. \qquad (17.60)$$

For the eigenvalue $-\kappa/N_a$ we get

$$x^2 \frac{d^2 u_2}{dx^2} + x \frac{du_2}{dx} + \left[-\frac{1}{4} x^2 + \left(\epsilon n_a + \frac{1}{2} \right) x - \gamma^2 |\kappa|^2 \right] u_2 = 0. \qquad (17.61)$$

Comparing to Eq. (16.38) we see that the Δ and Γ representations have comparable relations to Kramer's equation. In the Δ representation the solutions of Kramer's equation corresponding to $\pm\kappa/n_a$ are $u_1(x)$ and $u_2(x)$, which satisfy the differential equations (16.38). Solving these equations gives explicit expressions for $u_1(x)$ and $u_2(x)$. Finally, the radial components of the Dirac equation, $u_A(x)$ and $u_B(x)$ are obtained from $u_1(x)$ and $u_2(x)$ by using the known matrix of Eq. (16.18).

Notes on Chapter 17

Kramer's equation was used by Martin and Glauber [MG58] to obtain solutions of the Dirac equation, but also a similar idea was behind many of the methods discussed in the Notes on Chapter 15.

Chapter 18
Non-Relativistic Approximation

In this chapter we look at the non-relativistic limit of the theory developed in the previous chapters. For the energy levels we have of course the Bohr formula, as well as higher order effects. We are interested here in the reduction of the wave functions. It is of course well known that in the non-relativistic limit the Dirac wave functions reduce to the Pauli wave functions. Here we gain further insight into the separation of the Dirac radial functions into the two terms w_1 and w_2. It turns out that for negative values of κ the wave function u_A reduces to the two-component Pauli wave function u_{Nl}, and for positive values to the Pauli wave function $u_{N,l+1}$. Since the Dirac wave function is degenerate with respect to positive and negative values of κ, due to the supersymmetry, it is clear that both components u_{Nl} and $u_{N,l+1}$ will be present in the relativistic case. In the Δ-scheme, with its separation of the Dirac of the wave function into u_1 and u_2, we find the same result after the application of recursion relations of the hypergeometric functions.

The results of the computations in this chapter are not surprising. The explicit demonstration is nevertheless instructive, and verifies many of the calculations performed in the book. In particular we verify the choice of the overall phase, $Sign(\kappa)$, in Eq. (15.55), which ensures the equivalence of the expressions for the solutions of the radial Dirac equation, which were derived independently. Hence, our choice of phase is in coincidence with [DJ95].

Relativistic effects become important with increasing velocity. Let v be a typical velocity. For hydrogenic atoms the Bohr formula for the lowest state, $N = 1$, is

$$E_{Bohr} = E_N - m = -\frac{m}{2}(Z\alpha)^2$$

$$= \text{kinetic energy} + \text{potential energy} = -\text{kinetic energy}, \quad (18.1)$$

in accordance with the virial theorem [HGS01]: potential energy $= -2 \times$ kinetic energy. So the velocity v can be replaced by the more convenient parameter $\zeta = Z\alpha$.

With the formula (A.9):

$$(1 - z)^{-\frac{1}{2}} = \sum_{k=0}^{\infty} \frac{(1/2)_k}{k!} z^k = 1 + \frac{1}{2}z + \frac{3}{8}z^2 + O(z^3) \qquad (18.2)$$

we have for the Dirac energy

$$E = m \left[1 + \left(\frac{\zeta}{n_r + \gamma|\kappa|} \right)^2 \right]^{-\frac{1}{2}}$$

$$= m \left\{ 1 - \frac{\zeta^2}{2N^2} + \frac{\zeta^4}{N^3(2j+1)} + \frac{3}{8}\frac{\zeta^4}{N^4} \right\} + O(\zeta^6), \qquad (18.3)$$

with $N = n_r + |\kappa| = 1, 2, 3 \ldots, (N-1)/2$. To order ζ^2 we have, as expected,

$$\epsilon - 1 \approx \epsilon_{Bohr} = -\frac{\zeta^2}{2N^2}. \qquad (18.4)$$

The terms of order $O(\zeta^4)$ may be calculated in first-order perturbation theory from the Pauli formalism, for the relativistic mass-correction, the Thomas precession [Jac98], and the so-called Darwin term [Dar28], which is a result of the *zitterbewegung* [Ros61].

The apparent principal quantum number n_a has the non-relativistic approximation

$$n_a = \frac{\zeta}{\sqrt{1 - \epsilon^2}} \approx N. \qquad (18.5)$$

We then have

$$x \approx 2 \left(\frac{m\zeta}{N} \right) r. \qquad (18.6)$$

Note that $\gamma \approx 1 + O(\zeta^2)$. For positive κ we have

$$\kappa\epsilon - \gamma|\kappa| \approx \kappa - |\kappa| \approx O(\xi^2) \qquad (18.7)$$

and for negative κ, we have

$$\kappa\epsilon + \gamma|\kappa| \approx \kappa + |\kappa| \approx O(\xi^2). \qquad (18.8)$$

From Eq. (15.47) the radial quantum number is $n_r = \epsilon n_a - \gamma|\kappa| \approx N - |\kappa|$.

18.1 The Γ Representation

The exact expressions for the wave functions are, for the Γ-scheme:

$$u_A(x) = Sign(\kappa)\mathcal{N}x^{\gamma|\kappa|}e^{-x/2}\left[\sqrt{(\kappa) - \gamma|\kappa|)(\kappa\epsilon - \gamma|\kappa|)}\,_1F_1(-n_r, 2\gamma|\kappa|; x)\right.$$

$$\left. + \frac{en_a x}{2\gamma|\kappa|(2\gamma|\kappa| + 1)}\sqrt{(\kappa + \gamma|\kappa|)(\kappa\epsilon - \gamma|\kappa|)}\,_1F_1(-n_r + 1, 2\gamma|\kappa|+2; x)\right],$$

$$u_B(x) = \mathcal{N}x^{\gamma|\kappa|}e^{-x/2}\left[\sqrt{(\kappa + \gamma|\kappa|)(\kappa\epsilon - \gamma|\kappa|)}\,_1F_1(-n_r, 2\gamma|\kappa|; x)\right.$$

$$\left. + \frac{en_a x}{2\gamma|\kappa|(2\gamma|\kappa| + 1)}\sqrt{(\kappa - \gamma|\kappa|)(\kappa\epsilon + \gamma|\kappa|)}\,_1F_1(-n_r + 1, 2\gamma|\kappa|+2; x)\right]. \tag{18.9}$$

In the non-relativistic limit we have, from Eq. (D.16), for $|\kappa| = l$,

$$\mathcal{N} = \frac{1}{\Gamma(2\gamma|\kappa| + 1)}\sqrt{\frac{\Gamma(2\gamma|\kappa| + n_r)}{n_r!(2n_a)}} \approx \frac{1}{(2l)!}\sqrt{\frac{(N + l - 1)!}{(N - l)!(2N)}}. \tag{18.10}$$

For positive κ, the orbital angular momentum is $\bar{l} = j + \frac{1}{2} = |\kappa| = l + 1$, $\bar{n}_r = n_r - 1$, and $N = l + n_r = \bar{l} + \bar{n}_r$. We note that

$$en_a \approx \sqrt{N^2 - |\kappa|^2}. \tag{18.11}$$

The approximate expressions for the wave functions are

$$u_A(x) \approx \mathcal{N}x^{|\kappa|+1}e^{-x/2}\frac{\sqrt{N^2 - |\kappa|^2}}{(2|\kappa| + 1)}\,_1F_1(-n_r + 1, 2|\kappa| + 2; x)$$

$$= A_{N\bar{l}}x^{\bar{l}+1}e^{-x/2}\,_1F_1(-\bar{n}_r, 2\bar{l} + 2; x) = u_{N,l+1}(x), \tag{18.12}$$

since

$$\frac{1}{(2\bar{l})!(2\bar{l} + 1)}\sqrt{\frac{(N + \bar{l} - 1)!(N + \bar{l})(N - \bar{l})}{(N - \bar{l})!(2N)}}$$

$$= \frac{1}{(2\bar{l} + 1)!}\sqrt{\frac{(N + \bar{l})!}{(N - \bar{l} - 1)!(2N)}} = A_{N\bar{l}}, \tag{18.13}$$

from Eq. (9.29). The small component is $u_B(x) = O(\zeta)$.

For negative κ, the orbital angular momentum is $l = j - \frac{1}{2} = |\kappa| - 1$. The approximate expressions for the wave functions of Eq. (18.9) are then

$$u_A(x) \approx -2|\kappa|\mathcal{N}x^{|\kappa|}e^{-x/2}{}_1F_1(-n_r, 2|\kappa|; x)$$
$$= -A_{Nl}x^{l+1}e^{-x/2}{}_1F_1(-n_r, 2l+2; x) = -u_{Nl}(x). \quad (18.14)$$

where $N = n_r + |\kappa| = n_r + l + 1$. Here we have used

$$(2l+2)\mathcal{N} \approx \frac{1}{(2l+1)!}\sqrt{\frac{(N+l)!}{2N(N-l-1)!}} = A_{Nl}. \quad (18.15)$$

We have, again, $u_B(x) \approx O(\zeta)$.

The physical state is, for positive κ,

$$\psi \approx \begin{pmatrix} (1/r)u_{N,l+1}(r)\chi_\kappa(\theta, \phi) \\ O(\zeta) \end{pmatrix}, \quad (18.16)$$

and for negative κ,

$$\psi \approx \begin{pmatrix} -(1/r)u_{N,l}(r)\chi_\kappa(\theta, \phi) \\ O(\zeta) \end{pmatrix}. \quad (18.17)$$

In both cases the relativistic state reduces in the non-relativistic limit to the Pauli state.

18.2 The Δ Representation

For the Δ-scheme, the exact expressions for the wave functions are:

$$u_A(x) = \frac{\mathcal{N}'}{2}\sqrt{1+\epsilon}\, x^{\gamma|\kappa|}e^{-x/2}$$
$$\times [n_r{}_1F_1(-n_r+1, 2\gamma|\kappa|+1; x) - (n_a - \kappa){}_1F_1(-n_r, 2\gamma|\kappa|+1; x)],$$
$$u_B(x) = \frac{\mathcal{N}'}{2}\sqrt{1-\epsilon}\, x^{\gamma|\kappa|}e^{-x/2}$$
$$\times [n_r{}_1F_1(-n_r+1, 2\gamma|\kappa|+1; x) + (n_a - \kappa){}_1F_1(-n_r, 2\gamma|\kappa|+1; x)].$$
$$(18.18)$$

In the non-relativistic limit, for positive κ,

$$u_A(x) \approx \frac{\mathcal{N}'}{\sqrt{2}} x^{|\kappa|} e^{-x/2} n_r [_1F_1(-n_r + 1, 2|\kappa| + 1; x) - {}_1F_1(-n_r, 2|\kappa| + 1; x)],$$

$$u_B(x) \approx 0. \tag{18.19}$$

With the identities (E.4) and (D.4) we get

$$_1F_1(-n_r + 1, 2|\kappa| + 1; x) - {}_1F_1(-n_r, 2|\kappa| + 1; x)$$

$$= \left(\frac{x}{2|\kappa| + 1}\right) {}_1F_1(-n_r + 1, 2|\kappa| + 2; x). \tag{18.20}$$

Inserting this yields

$$u_A(x) \approx \frac{\mathcal{N}'}{\sqrt{2}} x^{|\kappa|+1} e^{-x/2} \left(\frac{n_r}{(2|\kappa| + 1)}\right) {}_1F_1(-n_r + 1, 2|\kappa| + 2; x). \tag{18.21}$$

Now use $|\kappa| = \bar{l}$ and $\bar{n}_r = n_r - 1$. From Eq. (E.12),

$$\mathcal{N}' = \frac{1}{\Gamma(2|\kappa| + 1)} \sqrt{\frac{\Gamma(n_r + 2|\kappa| + 1)}{n_a(n_a - \kappa) n_r!}} \approx \frac{1}{(2|\kappa|)!} \sqrt{\frac{(N + |\kappa|)!}{N \, n_r \, n_r!}} \tag{18.22}$$

and

$$\left(\frac{n_r}{2\bar{l} + 1}\right) \frac{1}{(2\bar{l})!} \sqrt{\frac{(N + \bar{l})!}{(2N) \, n, \, n_r!}} = \frac{1}{(2\bar{l} + 1)!} \sqrt{\frac{(N + \bar{l})!}{(2N) \, \bar{n}_r!}} = A_{N\bar{l}}. \tag{18.23}$$

Hence

$$u_A(x) \approx A_{N\bar{l}} x^{\bar{l}+1} e^{-x/2} {}_1F_1(-\bar{n}_r, 2\bar{l} + 1; x) = u_{N,l+1}(x). \tag{18.24}$$

For negative $-\kappa = |\kappa| = l + 1$ we have, in the non-relativistic limit,

$$u_A(x) \approx \frac{\mathcal{N}'}{\sqrt{2}} x^{|\kappa|} e^{-x/2}$$

$$\times [n_r {}_1F_1(-n_r + 1, 2|\kappa| + 1; x) - (n_r + 2|\kappa|) {}_1F_1(-n_r, 2|\kappa| + 1; x)]. \tag{18.25}$$

Use of the identity [GR65]

$$n_r {}_1F_1(-n_r + 1, 2|\kappa| + 1; x) - (n_r + 2|\kappa|) {}_1F_1(-n_r, 2|\kappa| + 1; x)$$

$$= -2|\kappa| {}_1F_1(-n_r, 2|\kappa|; x) \tag{18.26}$$

yields

$$u_A(x) \approx -\frac{\mathcal{N}'}{\sqrt{2}}\, x^{|\kappa|} e^{-x/2} 2|\kappa|\, {}_1F_1(-n_r, 2|\kappa|; x). \qquad (18.27)$$

Now use $|\kappa| = l + 1$.

$$u_A(x) \approx -\frac{\mathcal{N}'}{\sqrt{2}}\, x^{l+1} e^{-x/2}(2l+2)\, {}_1F_1(-N+l+1, 2l+2; x)$$

$$= -A_{Nl}x^{l+1}e^{-x/2}\, {}_1F_1(-N+l+1, 2l+2; x) = -u_{Nl}(x), \qquad (18.28)$$

since

$$(2l+2)\frac{\mathcal{N}'}{\sqrt{2}} \approx \frac{1}{(2l+1)!}\sqrt{\frac{(N+l+1)!}{2N\,(N+l+1)\,(N-l-1)!}}$$

$$= \frac{1}{2l+1}\sqrt{\frac{(N+l)!}{2N(N-l+1)!}} = A_{Nl}, \qquad (18.29)$$

from

$$\mathcal{N}' \approx \frac{1}{(2l+2)!}\sqrt{\frac{(N+l+1)!}{N(N+l+1)(N-l-1)!}}, \qquad (18.30)$$

see Eq. (E.12).

We have regained the wave functions of Eqs. (18.16) and (18.17).

Chapter 19
Conclusions

"There's a reason physicists are so successful at what they do, and that is they study the hydrogen atom and the helium ion and then they stop."
Attributed to Richard Feynman by J. Rigden, in *Hydrogen: The Essential Element*, Harvard University Press, Cambridge, MA, 2002.

One of the central concepts in this book is the Laplace vector, which gives us the connection between an attractive central force, which falls off inversely proportional to the distance squared, and the elliptical orbits that a particle moving under the influence of this force follows. It governs the motion of a planet about the Sun, and of an electron about the nucleus. It is valid in the classical domain, and, in a suitably modified form, also in the quantum domain. With a further generalization, a remnant of the Laplace vector even appears as a conserved quantity in the relativistic version of this problem at the quantum level.

Neglecting for a moment the spin of the electron, we see that the three conserved components of the Laplace vector $\mathcal{A}$, together with the three conserved components of the angular momentum vector L, suitably normalised, form a representation of the closed Lie algebra so(4). This gives a method for computing the spectrum of a non-relativistic hydrogen-like atom, including the degeneracy of the spectral lines.

In the classical problem the magnitude of the Laplace vector gives the eccentricity of the orbits. Sommerfeld constructed a semi-classical model of the hydrogen atom, which gives the correct positions of the spectral lines, and a characterization of the orbits according to their eccentricities. The model works for the non-relativistic hydrogen atom, as well as for the relativistic hydrogen atom. Even though it neglects the spin of the electron it gives the fine structure of the spectrum as it results from Dirac's relativistic theory, and the correct expression for the eccentricity of the orbits.

When we want to take the spin of the electron into account we are led naturally to the Pauli formalism, where the wave function of the electron has two components.

The electron momentum is replaced by the quantity $(\boldsymbol{\sigma} \cdot \mathbf{p})$, with σ_i, $i = 1, 2, 3$ the Pauli matrices. In a similar way the Laplace vector $\mathcal{A}$ is replaced by the quantity $A_{nr} = (\boldsymbol{\sigma} \cdot \mathcal{A})$, and the magnitude of this quantity is identified with the eccentricity: $e = \sqrt{A_{nr}^2}$. An important role is played by the eigenvalues of the operator $K_{nr} = -(\boldsymbol{\sigma} \cdot \mathbf{L} + 1)$, the non-relativistic counterpart of a constant operator introduced by Dirac. It turns out that the operator A_{nr} acts on eigenstates of K_{nr} in such a way as to change the eigenvalue κ of K_{nr} into $-\kappa$. In this way it connects two states of the system with identical energies, and establishes a *supersymmetry* of the system. It allows us to construct Hermitian conjugate operators A_- and A_+ which act as ladder operators to construct all states of the system from a supersymmetric vacuum. The complete degeneracy of the spectrum is accounted for.

The Dirac equation is introduced as a straightforward generalization of the Pauli equation to four-component spinors, with the γ_μ matrices, $\mu = 1$ to 4, replacing the Pauli matrices. The primary supersymmetry of the states of an electron in a Coulomb field is generated by the Johnson–Lippmann operator A, the relativistic version of the operator A_{nr}, to which it reduces in the non-relativistic limit. The SO(4) symmetry is lost, as is the SO(3) symmetry associated to the orbital angular momentum. But there is an SO(3) symmetry associated to the total angular momentum, including the spin. Thus the (super)symmetry of the relativistic hydrogen atom is SO(3) × S(2).

Recognizing the supersymmetry of the Dirac equation allows us to construct in a systematic way its solutions. It turns out that extending the space of solutions to include the solutions of the Dirac equation for negative mass leads to a pair of confluent hypergeometric functions $w_1(x)$ and $w_2(x)$, of which definite linear combinations form solutions of the Dirac equation. These hypergeometric functions are solutions of second-order differential equations that factorize under supersymmetry to yield first-order equations involving conjugate operators a_- and a_+ that are relativistic generalizations of the operators A_- and A_+ used in the Pauli formalism.

Another extension of the space of solutions leads to an alternative pair of confluent hypergeometric functions $u_1(x)$ and $u_2(x)$, which are known forms of the solutions of the Dirac equation.

An alternative method of solving the Dirac equation consists of solving a second-order differential equation of Schrödinger form, the Kramer's equation. Its solutions $w_1(x)$ and $w_2(x)$ live in the extended solution space described above. It is then necessary to project out from these solutions the appropriate linear combinations that are solutions of the Dirac equation. An analysis of this method reveals that it again involves a factorization of the second-order differential operator identical to that discussed above.

Finally, performing the non-relativistic reduction of the solutions of the Dirac equation is shown to lead to the well-known solutions of the Pauli equation. This serves as a consistency check of the normalizations and the phase choices made in the course of developing the formalism.

Feynman's comment on the successes achieved through the methods employed here is a measure of the extent to which relativistic quantum mechanics is amenable to exact solution. Further investigations of electromagnetic processes are all limited to studying perturbative expansions based on the exact solutions.

Figure 19.1 attempts to show how our understanding of the spectrum of hydrogen has developed over the years, from the Bohr model for the non-relativistic case, through the relativistic Sommerfeld model and the Dirac theory, up to the Lamb shift, which demonstrates the need for a full quantum field-theoretic treatment.

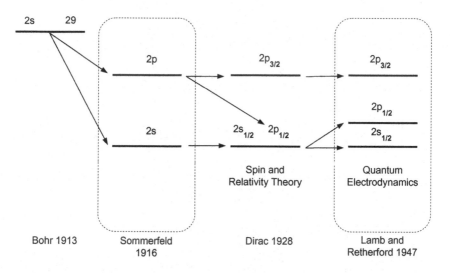

Figure 19.1. The development of our understanding of the hydrogen spectrum.

Appendices

Appendix A. The Confluent Hypergeometric Function

The hypergeometric functions are a wide class that include all the special functions of mathematical physics, such as the Hermite, Bessel and Legendre polynomials, and the associated Legendre polynomials. The Laguerre polynomials and the associated Laguerre functions are described by the *confluent hypergeometric function*.

The confluent hypergeometric function is a solution of the differential equation

$$\left(x\frac{d^2}{dx^2} + (c-x)\frac{d}{dx} - a \right) {}_1F_1(a, c; x) = 0. \tag{A.1}$$

It has solutions

$$ {}_1F_1(a, c; x) = a_0 \sum_{k=0}^{\infty} \frac{(a)_k}{k!(c)_k} x^k, \tag{A.2}$$

where $(a)_k$ are the *Pochhammer symbols* defined by

$$(a)_0 = 1,$$
$$(a)_n = a(a+1)(a+2)\cdots(a+n-1), \quad n = 1, 2, 3, \ldots. \tag{A.3}$$

They satisfy identities such as

$$n! = (n-m)!(n-m+1)_m,$$
$$(c-m+1)_m = (-1)^m(-c)_m, \tag{A.4}$$
$$(n+m)! = n!(n+1)_m.$$

A function that is analytic in the interval $a < x < b$ may be expanded in a Taylor series

$$f(x) = \sum_{k=0}^{\infty} \frac{f^{(n)}(x_0)}{n!} (x - x_0)^n, \tag{A.5}$$

for any x_0 in the interval.

For example, the function

$$f(z) = (1 - z)^s \tag{A.6}$$

is analytic for any real s with z in the interval $-1 < z < 1$. Its derivatives are

$$
\begin{aligned}
f'(z) &= -s(1 - z)^{s-1} \\
f''(z) &= (-1)^2 s(s - 1)(1 - z)^{s-2} \\
f^{(k)}(z) &= (-1)^k s(s - 1) \cdots (s - k + 1)(1 - z)^{s-k},
\end{aligned} \tag{A.7}
$$

from which we can construct the Taylor series,

$$(1 - z)^s = \sum_{k=0}^{\infty} \frac{[f^{(k)}(z)]_{z=0}}{k!} z^k = \sum_{k=0}^{\infty} \frac{(-1)^k s(s - 1) \cdots (s - k + 1)}{k!} z^k. \tag{A.8}$$

This can be written

$$(1 - z)^s = \sum_{k=0}^{\infty} \frac{(-s)_k}{k!} z^k, \tag{A.9}$$

where we have used the identity $(-s)_k = (-1)^k (s - k + 1)_k$. With $s = -1$, this reduces to the *geometric series*

$$\frac{1}{1 - z} = \sum_{k=0}^{\infty} z^k, \tag{A.10}$$

which converges for $|z| < 1$.

An important identity of the Pochhammer symbols is given by *Vandermonde's theorem*:

$$\sum_{m=0}^{n} \frac{(a)_m}{m!} \frac{(b)_{n-m}}{(n - m)!} = \frac{(a + b)_n}{n!}, \tag{A.11}$$

where a and b are numbers independent of the summation index m. To prove it we express s in Eq. (A.9) as the sum of two numbers a and b and set $z = -x$ to obtain

$$(1 + x)^{a+b} = \sum_{n=0}^{\infty} \frac{(a + b - n + 1)_n}{n!} x^n. \tag{A.12}$$

This function can also be written as a product

$$(1+x)^{a+b} = (1+x)^a (1+x)^b = \sum_{k=0}^{\infty} \frac{(a-k+1)_k}{k!} x^k \sum_{j=0}^{\infty} \frac{(b-j+1)_j}{j!} x^j.$$

(A.13)

This is the Cauchy product of two series

$$\sum_{k=0}^{\infty} S_k \sum_{j=0}^{\infty} T_j = \sum_{n=0}^{\infty} U_n,$$

(A.14)

where

$$U_n = \sum_{m=0}^{n} S_m T_{n-m}.$$

(A.15)

So the Cauchy product of the two series in Eq. (A.13) is

$$(1+x)^{a+b} = \sum_{n=0}^{\infty} \sum_{m=0}^{n} \frac{(a-m+1)_m}{m!} \frac{(b-n+m+1)_{n-m}}{(n-m)!} x^n.$$

(A.16)

Subtraction of Eq. (A.16) from Eq. (A.12) yields

$$\sum_{n=0}^{\infty} \left[\frac{(a+b-n+1)_n}{n!} - \sum_{m=0}^{n} \frac{(a-m+1)_m}{m!} \frac{(b-n+m+1)_{n-m}}{(n-m)!} \right] x^n = 0.$$

(A.17)

Since this holds for arbitrary x in the interval $-1 < x < 1$, the terms in different powers of x are linearly independent. Therefore, the coefficient of each power of x must vanish separately, that is,

$$\sum_{m=0}^{n} \frac{(a-m+1)_m}{m!} \frac{(b-n+m+1)_{n-m}}{(n-m)!} = \frac{(a+b-n+1)_n}{n!}.$$

(A.18)

Now use Eq. (A.4) to get

$$\sum_{m=0}^{\infty} \frac{(-a)_m}{m!} \frac{(-b)_{n-m}}{(n-m)!} = \frac{(-a-b)_n}{n!}.$$

(A.19)

On replacing $-a$ by a and $-b$ by b, Eq. (A.11) follows.

Laguerre polynomials

Consider the function $_1F_1(-q, 1; x)$. To have a polynomial q must be a positive integer. Then, in the series

$$_1F_1(-q, 1; x) = \sum_{k=0} \frac{(-q)_k}{k!k!} x^k \qquad (A.20)$$

all terms for $k > q$ vanish. Now,

$$\frac{(-q)_k}{k!} = (-1)^k \frac{(q-k+1)_k}{k!} = (-1)^k \sum_{m=0}^{k} \frac{(q+1)_m}{m!} \frac{(-k)_{k-m}}{(k-m)!}, \qquad (A.21)$$

where we have invoked Vandermonde's theorem.

With this result we can write,

$$_1F_1(-q, 1; x) = \sum_{k=0}^{\infty} \frac{(-1)^k}{k!} \left(\sum_{m=0}^{\infty} \frac{(q+1)_m}{m!} \frac{(-k)_{k-m}}{(k-m)!} \right) x^k. \qquad (A.22)$$

Now we use the identities

$$(m+q)! = q!(q+1)_m \qquad (A.23)$$

and

$$k! = (-1)^{k-m} m! (-k)_{k-m}, \quad m \le k, \qquad (A.24)$$

to obtain

$$_1F_1(-q, 1; x)$$

$$= \frac{1}{q!} \sum_{k=0}^{\infty} \left(\sum_{m=0}^{k} \frac{1}{(k-m)!} x^{k-m} \frac{(-1)^m (m+1) \cdots (m+q)}{m!} x^m \right)$$

$$= \frac{1}{q!} \sum_{k=0}^{\infty} \left(\sum_{m=0}^{k} \frac{x^{k-m}}{(k-m)!} \frac{(-1)^m}{m!} \frac{d^q}{dx^q} x^{m+q} \right). \qquad (A.25)$$

We recognize this as the Cauchy product of two series. Thus,

$$_1F_1(-q, 1; x) = \frac{1}{q!} \sum_{r=0}^{\infty} \frac{x^r}{r!} \frac{d^q}{dx^q} \sum_{s=0}^{\infty} \frac{(-1)^s}{s!} x^{s+q} = \frac{1}{q!} e^x \frac{d^q}{dx^q} (x^q e^x), \qquad (A.26)$$

which is a polynomial of degree q. It is, up to a normalization constant, the *Laguerre polynmial* of Eq. (9.35).

The generating function for the Laguerre polynomials is

$$\frac{e^{-xt/(1-t)}}{1-t} = \sum_{k=0}^{\infty} \frac{L_k(x)}{k!} t^k. \tag{A.27}$$

Let us differentiate both sides of Eq. (A.27) p times with respect to x,

$$\sum_{k=0}^{\infty} \frac{1}{k!} \left[\frac{d^p}{dx^p} L_k(x) \right] t^k = \sum_{k=0}^{\infty} \frac{L_k^p(x)}{k!} t^k. \tag{A.28}$$

If this differentiation is carried out in Eq. (A.27), the generating function for the associated Laguerre polynomials is found to be

$$(-1)^p \frac{t^p}{(1-t)^{p+1}} e^{-xt/(1-t)} = \sum_{q=0}^{\infty} \frac{L_q^p(x)}{q!} t^q. \tag{A.29}$$

Notes on Appendix A

A good source of information on hypergeometric functions is the book by Seaborn [Sea80]. The recurrence relations used in the text are given there in terms of the associated Laguerre polynomials.

**Appendix B. Orthogonality Relations
of Hypergeometric Functions**

We wish to calculate the normalization of the radial eigenfunctions of the bound states of a particle in a Coulomb potential, Eq. (9.39):

$$R_{Nl}(r) = \left(\frac{2Z\alpha m}{N} \right)^{\frac{3}{2}} N_{Nl} \, x^l \, e^{-\frac{1}{2}x} \, L_{N+l}^{2l+1}(x), \tag{B.1}$$

with $x = 2(Z\alpha m/N)r$. The complete eigenfunctions $R_{Nl}(r)Y_l^m(\theta, \phi)$ are considered to be orthonormal; therefore

$$\int_0^{\infty} R_{Nl}(r) R_{N'l}(r) r^2 \, dr = \delta_{NN'}. \tag{B.2}$$

This leads to the appropriate orthogonality relation for the Laguerre polynomials. By substituting from Eq. (B.1) into the integral in Eq. (B.2) and changing the variable to x we find that

$$\int_0^\infty [R_{Nl}(r)]^2 r^2 \, dr = N_{Nl}^2 \int_0^\infty x^{2l+2} e^{-x} [L_{N+1}^{2l+1}(x)]^2 \, dx. \qquad (B.3)$$

Clearly, we are interested in the integral

$$\int_0^\infty x^{p+1} e^{-x} [L_q^p(x)]^2 \, dx. \qquad (B.4)$$

The form of the integrand in Eq. (B.4) suggests using the generating function in Eq. (A.29) to write

$$x^{p+1} e^{-x} \left[(-1)^p \frac{u^p}{(1-u)^{p+1}} e^{-xu/(1-u)} \right] \left[(-1)^p \frac{v^p}{(1-v)^{p+1}} e^{-xv/(1-v)} \right]$$

$$= \sum_{j=0}^\infty \sum_{k=0}^\infty \frac{u^j v^k}{j! k!} x^{p+1} e^{-x} L_j^p(x) L_k^p(x). \qquad (B.5)$$

We integrate both sides over the full range of x to obtain

$$(1-u)(1-v) \frac{\Gamma(p+2)(uv)^p}{(1-uv)^{p+2}} = \sum_{j=0}^\infty \sum_{k=0}^\infty \frac{u^j v^k}{j! k!} L_j^p(x) L_k^p(x). \qquad (B.6)$$

In carrying out this integration we use

$$\int_0^\infty x^z e^{-ax} dx = \frac{\Gamma(z+1)}{a^{z+1}} \qquad (B.7)$$

to show that

$$\int_0^\infty x^{p+1} e^{-x[1+u/(1-u)+v/(1-v)]} dx$$

$$= \Gamma(p+2) \left[1 + \frac{u}{(1-u)} + \frac{v}{(1-v)} \right]^{-p-2}$$

$$= \Gamma(p+2) \left[\frac{(1-u)(1-v)}{(1-uv)} \right]^{p+2}. \qquad (B.8)$$

We now use Eqs. (A.5) and (A.10) in

$$(1-uv)^{-(p+2)} = \sum_{k=0}^\infty \frac{(p+2)_k}{k!} (uv)^k \qquad (B.9)$$

to obtain

$$(1-u)(1-v)(p+1)! \sum_{k=0}^{\infty} \frac{(p+2)_k}{k!}(uv)^{k+p}$$

$$= (1-u)(1-v) \sum_{k=0}^{\infty} \frac{(p+k+1)!}{k!}(uv)^{k+p}$$

$$= \int_0^{\infty} L_j^p(x) L_k^p(x) dx. \tag{B.10}$$

Since u and v are arbitrary and independent, the coefficients of $u^j v^k$ for any given j and k must be the same on both sides of this equation. In particular if $j = k = q$,

$$\frac{(q+1)!}{(q-p)!} + \frac{q!}{(q-p-1)!} = \frac{1}{(q!)^2} \int_0^{\infty} x^{p+1} e^{-x} [L_q^p(x)]^2 dx. \tag{B.11}$$

From this result the value of the integral is seen to be

$$\int_0^{\infty} x^{p+1} e^{-x} [L_q^p(x)]^2 dx = \frac{[\Gamma(q+1)]^3}{(q-p)!}(2q+1-p). \tag{B.12}$$

Another integral we will need for normalizing the Dirac radial functions is

$$\int_0^{\infty} x^p e^{-x} [L_q^p(x)]^2 \, dx. \tag{B.13}$$

Following the same procedure as before we get

$$x^p e^{-x} \left[(-1)^p \frac{u^p}{(1-u)^{p+1}} e^{-\frac{xu}{(1-u)}} \right] \left[(-1)^p \frac{v^p}{(1-v)^{p+1}} e^{-\frac{xv}{(1-v)}} \right]$$

$$= \sum_{j=0}^{\infty} \sum_{k=0}^{\infty} \frac{u^j v^k}{j!k!} x^p e^{-x} L_j^p(x) L_k^p(x). \tag{B.14}$$

We integrate both sides over the full range of x to obtain

$$\frac{\Gamma(p+1)(uv)^p}{(1-uv)^{p+1}} = \sum_{j=0}^{\infty} \sum_{k=0}^{\infty} \frac{u^j v^k}{j!k!} \int_0^{\infty} x^p e^{-x} L_j^p(x) L_k^p(x) dx. \tag{B.15}$$

In carrying out this integration we this time get

$$\int_0^{\infty} x^p e^{-x[1+u/(1-u)+v/(1-v)]} dx = \Gamma(p+1) \left[1 + \frac{u}{(1-u)} + \frac{v}{(1-v)} \right]^{-p-1}. \tag{B.16}$$

If the denominator on the left-hand side of Eq. (B.15) is expanded as in Eq. (A.9), then Eq. (B.15) becomes

$$\Gamma(p+1)\sum_{s=0}^{\infty}\frac{(p+1)_s}{s!}(uv)^{s+p} = \sum_{j=0}^{\infty}\sum_{k=0}^{\infty}\frac{u^j v^k}{j!k!}\int_0^{\infty} x^p e^{-x} L_j^p(x)L_k^p(x)dx.$$

(B.17)

Now, $L_j^p(x) = 0$ for $j < p$, therefore, we can change the lower limits on the sums on the right-hand side of Eq. (B.17) to p without affecting the sums. Next, redefine the summation index on the left-hand side to be $j = s + p$. This gives

$$\Gamma(p+1)\sum_{j=p}^{\infty}\frac{(p+1)_{j-p}}{(j-p)!}(uv)^j = \sum_{j=p}^{\infty}\sum_{k=p}^{\infty}\frac{u^j v^k}{j!k!}\int_0^{\infty} x^p e^{-x} L_j^p(x)L_k^p(x)dx.$$

(B.18)

Now use the definition of the Kronecker delta to rewrite the left-hand side of this equation. After some rearrangement of terms the equation then reads

$$\sum_{j=p}^{\infty}\sum_{k=p}^{\infty} u^j v^k \left[\frac{1}{j!k!}\int_0^{\infty} x^p e^{-x} L_j^p(x)L_k^p(x)dx\right.$$
$$\left. - \frac{\Gamma(p+1)(p+1)_{k-p}}{(k-p)!}\delta_{jk}\right] = 0.$$

(B.19)

The coefficients of $u^j v^k$ for different combinations of j and k must vanish separately. Thus,

$$\int_0^{\infty} x^p e^{-x} L_j^p(x)L_k^p(x)dx = \frac{(k!)^2 p!(p+1)_{k-p}}{(k-p)!}\delta_{jk}.$$

(B.20)

With $k - p$ equal to an integer $\Gamma(p+1)(p+1)_{k-p} = \Gamma(k+1)$, from Eq. (A.4). Therefore, for the more general case where $q - p$ is an integer, but q and p are not necessarily integers,

$$\int_0^{\infty} x^p e^{-x} [L_q^p(x)]^2\, dx = \frac{[\Gamma(q+1)]^3}{(q-p)!}.$$

(B.21)

Expressed in terms of $_1F_1(-n; 2a; x)$ the orthogonality relation for the $L_n^p(x)$ becomes

$$\int_0^{\infty} e^{-x} x^{2a}\, _1F_1^2(-n, 2a; x)dx = \frac{[\Gamma(2a+1)]^2 n!^2}{[\Gamma(2a+n)]^4}\int_0^{\infty} e^{-x} x^{2a}[L_{2a+n}^{2a}(x)]^2 dx$$

$$= 2(a+n)\frac{[\Gamma(2a)]^2}{\Gamma(2a+n)}n!,$$

(B.22)

where we have used Eq. (9.38).

From (B.12) we see that the normalized eigenfunctions are

$$R_{Nl}(r) = -\sqrt{\frac{(N-l-1)!}{(N+l)!^3(2N)}} \left(\frac{2Z\alpha m}{N}\right)^{\frac{3}{2}} e^{-Z\alpha mr/N}$$

$$\times \left(\frac{2Z\alpha mr}{N}\right)^l L_{N+1}^{2l+1}(2Z\alpha r/N)$$

$$= \frac{1}{(2l+1)!}\sqrt{\frac{(N+l)!}{(N-l-1)!(2N)}} \left(\frac{2Z\alpha m}{N}\right)^{\frac{3}{2}} e^{-(Z\alpha mr/N)}$$

$$\times \left(\frac{2Z\alpha mr}{N}\right)^l {}_1F_1(-N+l+1, 2l+2; 2Z\alpha r/N). \qquad \text{(B.23)}$$

Notes on Appendix B

Be careful of different conventions concerning the associated Laguerre functions, for example Mertzbacher [Mer61] vs Landau and Lipshitz [LL77].

Appendix C. More Integrals Involving Hypergeometric Functions

For the sake of completeness, we give here the application of the formulas of Landau and Lipshitz [LL77], Appendix F, dealing with the calculation of integrals involving confluent hypergeometric functions.

To calculate the normalization we need the following integrals:

$$I_{11} = \int_0^\infty z^{2a+2} e^{-z} [{}_1F_1(1-n, 2a+2; z)]^2 \, dz,$$

$$I_{22} = \int_0^\infty z^{2a} e^{-z} [{}_1F_1(-n, 2a; z)]^2 \, dz, \qquad \text{(C.1)}$$

$$I_{12} = \int_0^\infty z^{2a+1} e^{-z} {}_1F_1(-n, 2a; z) {}_1F_1(1-n, 2a+2; z) \, dz.$$

These integrals are special cases of the integral (F.16) discussed by Landau and Lipshitz, namely

$$J_\mu^{sp}(\alpha, \alpha') = \int_0^\infty z^{\mu-1+s} e^{-(k+k')z/2} {}_1F_1(\alpha, \mu; kz) {}_1F_1(\alpha'; \mu-p; k'z) dz.$$

$$\text{(C.2)}$$

To calculate these integrals we make use of the hypergeometric function

$$_2F_1(a, b, c; z) = \sum_{n=0}^{\infty} \frac{(a)_n (b)_n}{n!(c)_n} z^n. \tag{C.3}$$

The definition of the generalized hypergeometric function is

$$_pF_q(a_1, a_2, \ldots, a_p, b_1, b_2, \ldots, b_q; z) = \sum_{n=0}^{\infty} \frac{(a_1)_n (a_2)_n \cdots (a_p)_n}{n!(b_1)_n (b_2)_n \cdots (b_q)_n} z^n, \tag{C.4}$$

where p is the number of numerator parameters a_i and q is the number of denominator parameters b_j.

To specialize (C.2) to the integrals above, we must set $k = k'$, which we do by choosing $k = 1 + \epsilon$, $k' = 1 - \epsilon$ and then taking the limit $\epsilon \to 0$. The need for this procedure is made clear by looking at a particular case of (C.2), namely, $s = p = 0$, which, according to (F.13), is

$$J_\mu^{00}(\alpha, \alpha') = (-1)^\alpha \Gamma(\mu) \epsilon^{-(\alpha + \alpha')} {}_2F_1(\alpha, \alpha', \mu; -\epsilon^{-2}), \tag{C.5}$$

where we have used the hypergeometric function

$$_2F_1(\alpha, \alpha', \mu; z) = 1 + \frac{\alpha\alpha'}{\mu} \frac{z}{1!} + \frac{\alpha(\alpha + 1)\alpha'(\alpha' + 1)}{\mu(\mu + 1)} \frac{z^2}{2!} + \cdots. \tag{C.6}$$

The indices α and α' are negative integers, so that Eq. (C.6) is a polynomial which, in the limit $\epsilon \to 0$, is determined by the last term only. This is summarized in the relation

$$\epsilon^m {}_2F_1(\alpha, \alpha', \mu; -\epsilon^{-2}) \xrightarrow[\epsilon \to 0]{} \begin{cases} 0 & m > 2n_2, \\ (-1)^{n_2} \dfrac{n_1!}{(n_1 - n_2)!} \dfrac{\Gamma(\mu)}{\Gamma(\mu + n_2)} & m = 2n_2 \\ \text{undefined} & m < 2n_2 \end{cases} \tag{C.7}$$

where $n_1 = \max(-\alpha, -\alpha')$, and $n_2 = \min(-\alpha, -\alpha')$.

The integrals in (C.2) are

$$I_{11} = J_{2a+2}^{10}(1 - n, 1 - n),$$
$$I_{22} = J_{2a}^{10}(-n, -n), \tag{C.8}$$
$$I_{12} = J_{2a+2}^{02}(1 - n, n).$$

The relationship (F.16), for $p = 0$,

$$J_\mu^{10}(\alpha, \alpha') = \left(\frac{\alpha' - \alpha}{\epsilon} + \mu - \alpha' - \alpha\right) J_\mu^{00}(\alpha, \alpha'), \qquad (C.9)$$

enables us to evaluate the first two integrals of (C.9):

$$I_{11} = (2a + 2n)J_{2a+2}^{00}(1 - n, 1 - n)$$

$$= 2(a + n)\frac{[\Gamma(2a + 2)]^2}{\Gamma(2a + n + 1)}(n - 1)!$$

$$I_{22} = (2a + 2n)J_{2a}^{00}(-n, -n) = 2(a + n)\frac{[\Gamma(2a)]^2}{\Gamma(2a + n)}n! \qquad (C.10)$$

The value of I_{22} calculated here agrees with Eq. (B.22), calculated by more elementary methods.

A little more work is required for I_{12}. The recursion formula, Eq. (F.15), namely

$$J_\mu^{sp}(\alpha, \alpha') = (\mu - 1)[J_{\mu-1}^{s,p-1}(\alpha, \alpha') - J_{\mu-1}^{s,p-1}(\alpha - 1, \alpha')], \qquad (C.11)$$

applied twice, yields

$$J_{2a+2}^{02}(1 - n, -n) = 2a(2a + 1)[J_{2a}^{00}(1 - n, -n))$$
$$- 2J_{2a}^{00}(-n, -n) + J_{2a}^{00}(-1 - n, -n)]. \qquad (C.12)$$

According to (C.7), only the middle term survives, and

$$I_{12} = -2(2a + 1)\frac{[\Gamma(2a)]^2}{\Gamma(2a + n)}\, n! \qquad (C.13)$$

Notes on Appendix C

Appendix F of Landau and Lipshitz [LL77] is an invaluable source of information on integrals of hypergeometric functions, which often seems to be insufficiently appreciated. The limit processes necessary for the application of the results of this appendix to the cases considered follow [GI97].

Appendix D. Normalization for the Γ-Induced Scheme

We have the solutions

$$w_1(x) = N_1 x^{\gamma|\kappa|} e^{-x/2} {}_1F_1(-n_r, 2\gamma|\kappa|; x)$$
$$w_2(x) = N_2 x^{\gamma|\kappa|+1} e^{-x/2} {}_1F_1(-n_r + 1, 2\gamma|\kappa| + 2; x). \tag{D.1}$$

To determine N_1/N_2 insert this into the first of the equations in (15.34), where b^- is defined in Eq. (15.33):

$$b^- w_1 = \left[-\frac{d}{dx} + \left(\frac{\gamma|\kappa|}{x} - \frac{\epsilon n_a}{2\gamma|\kappa|} \right) \right] w_1 = -\frac{\epsilon n_a}{2\gamma|\kappa|} w_2. \tag{D.2}$$

First calculate

$$\frac{dw_1}{dx} = N_1 x^{\gamma|\kappa|} e^{-x/2} \left\{ \left(\frac{\gamma|\kappa|}{x} - \frac{1}{2} \right) {}_1F_1(-n_r, 2\gamma|\kappa|; x) \right.$$
$$\left. + \frac{d_1 F_1(-n_r, 2\gamma|\kappa|; x)}{dx} \right\}. \tag{D.3}$$

Use the relation

$$\frac{d_1 F_1(-n_r, 2\gamma|\kappa|; x)}{dx} = \frac{-n_r}{2\gamma|\kappa|} {}_1F_1(-n_r + 1, 2\gamma|\kappa| + 1; x). \tag{D.4}$$

Eq. (D.2) then becomes

$$N_1 \left\{ \left(\frac{\epsilon n_a}{2\gamma|\kappa|} - \frac{1}{2} \right) {}_1F_1(-n_r, 2\gamma|\kappa|; x) \right.$$
$$\left. - \left(\frac{n_r}{2\gamma|\kappa|} \right) {}_1F_1(-n_r + 1, 2\gamma|\kappa| + 1; x) \right\}$$
$$= N_2 \left(\frac{\epsilon n_a}{2\gamma|\kappa|} \right) x_1 F_1(-n_r + 1, 2\gamma|\kappa| + 2; x). \tag{D.5}$$

We remember that $n_r = \epsilon n_a - \gamma|\kappa|$. Then

$$\left(\frac{n_r}{2\gamma|\kappa|} \right) N_1 \left\{ {}_1F_1(-n_r, 2\gamma; x) - {}_1F_1(-n_r + 1, 2\gamma|\kappa| + 1; x) \right\}$$
$$= N_2 \left(\frac{\epsilon n_a}{2\gamma|\kappa|} \right) x_1 F_1(-n_r + 1, 2\gamma|\kappa| + 2; x). \tag{D.6}$$

Use the relation [GR65]

$$-\frac{(n_r + 2\gamma|\kappa|)}{2\gamma|\kappa|(2\gamma|\kappa| + 1)}x_1F_1(-n_r + 1, 2\gamma|\kappa| + 2; x)$$

$$= {}_1F_1(-n_r, 2\gamma|\kappa|; x) - {}_1F_1(-n_r + 1, 2\gamma|\kappa| + 1; x). \qquad (D.7)$$

This yields

$$N_1\left(\frac{n_r}{2\gamma|\kappa|}\right)\frac{(n_r + 2\gamma|\kappa|)}{2\gamma|\kappa|(2\gamma|\kappa| + 1)}x_1F_1(-n_r + 1, 2\gamma|\kappa| + 2; x)$$

$$= -N_2\left(\frac{en_a}{2\gamma|\kappa|}\right)x_1F_1(-n_r + 1, 2\gamma|\kappa| + 2; x). \qquad (D.8)$$

The result is

$$\frac{N_1}{N_2} = -\left(\frac{en_a}{n_r}\right)\frac{2\gamma|\kappa|(2\gamma|\kappa| + 1)}{(n_r + 2\gamma|\kappa|)} = -\left(\frac{1}{en_a}\right)2\gamma|\kappa|(2\gamma|\kappa| + 1), \qquad (D.9)$$

with $(n_r + 2\gamma|\kappa|)n_r = (en_a + \gamma|\kappa|)(en_a - \gamma|\kappa|) = \epsilon^2 n_a^2 - \gamma^2|\kappa|^2 = n_a^2 - \kappa^2 = (en_a)^2$, and we have used Eqs. (14.33), (13.61) and (14.50).

We now wish to establish the overall normalization constant. We have the normalization condition

$$\int \left[u_A^2(x) + u_B^2(x)\right] dx = 1. \qquad (D.10)$$

Inserting $u_A(x)$ and $u_B(x)$ yields

$$\mathcal{N}^2\left\{2\kappa(\kappa\epsilon - \gamma|\kappa|)\int dx\, x^{2\gamma|\kappa|}e^{-x}[{}_1F_1(-n_r, 2\gamma|\kappa|; x)]^2\right.$$

$$+ \frac{4en_a}{2\gamma|\kappa|(2\gamma|\kappa| + 1)}\sqrt{(\kappa^2 - \gamma^2|\kappa|^2)(\kappa^2\epsilon^2 - \gamma^2|\kappa|^2)}$$

$$\times \int dx\, x^{2\gamma|\kappa|+1}e^{-x}{}_1F_1(-n_r, 2\gamma|\kappa|; x)_1F_1(-n_r + 1, 2\gamma|\kappa| + 2; x)$$

$$+ \frac{(en_a)^2}{(2\gamma|\kappa|)^2(2\gamma|\kappa| + 1)^2}2\kappa(\kappa\epsilon + \gamma|\kappa|)$$

$$\left.\times \int dx\, x^{2\gamma|\kappa|+2}e^{-x}[{}_1F_1(-n_r + 1, 2\gamma|\kappa| + 2 : x)]^2\right\} = 1. \qquad (D.11)$$

Evaluating the integrals according to Appendix C:

$$\int x^{2\gamma|\kappa|}e^{-x}[_1F_1(-n_r,2\gamma|\kappa|;x)]^2\,dx = 2(2\gamma|\kappa|+n_r)\frac{[\,\Gamma(2\gamma|\kappa|)\,]^2}{\Gamma(2\gamma|\kappa|+n_r)}n_r!$$

$$\int x^{2\gamma|\kappa|+2}e^{-x}[_1F_1(-n_r+1,2\gamma|\kappa|+2;x)]^2\,dx$$

$$= 2(2\gamma|\kappa|+n_r)\frac{[\Gamma(2\gamma|\kappa|+2)]^2}{\Gamma(2\gamma|\kappa|+n_r+1)}(n_r-1)!$$

$$\int x^{2\gamma|\kappa|+1}e^{-x}\,_1F_1(-n_r,2\gamma|\kappa|;x)_1F_1(-n_r+1,2\gamma|\kappa|+2;x)dx$$

$$= -2(2\gamma|\kappa|)(2\gamma|\kappa|+1)\frac{[\Gamma(2\gamma|\kappa|)]^2}{\Gamma(2\gamma|\kappa|+n_r)}(n_r)! \tag{D.12}$$

so that

$$\frac{1}{\mathcal{N}^2} = 4\frac{[\,\Gamma(2\gamma|\kappa|)\,]^2}{\Gamma(2\gamma|\kappa|+n_r)}\,n_r!$$

$$\left\{\kappa(\kappa\epsilon-\gamma|\kappa|)\epsilon n_a - 2(en_a)Z\alpha|\kappa|\sqrt{\epsilon^2-\gamma^2}\right.$$

$$\left. + \frac{\kappa(\kappa\epsilon+\gamma|\kappa|)(en_a)^2\epsilon n_a}{n_r(n_r+2\gamma|\kappa|)}\right\}. \tag{D.13}$$

Now use

$$n_r(n_r+2\gamma|\kappa|) = (en_a)^2 = \frac{|\kappa|^2(\epsilon^2-\gamma^2)}{1-\epsilon^2} \tag{D.14}$$

to reduce this to

$$\frac{1}{\mathcal{N}^2} = \frac{8n_a[\,\Gamma(2\gamma|\kappa|)\,]^2}{\Gamma(2\gamma|\kappa|+n_r)}\gamma^2|\kappa|^2\,n_r!, \tag{D.15}$$

which is equivalent to

$$\mathcal{N} = \frac{1}{\Gamma(2\gamma|\kappa|+1)}\sqrt{\frac{\Gamma(2\gamma|\kappa|+n_r)}{(2n_a)n_r!}}. \tag{D.16}$$

Appendix E. Normalization for the Δ-Induced Scheme

We have the solutions

$$u_1(x) = N_1 x^{\gamma|\kappa|} e^{-x/2} {}_1F_1(-n_r + 1, 2\gamma|\kappa| + 1; x)$$
$$u_2(x) = N_2 x^{\gamma|\kappa|} e^{-x/2} {}_1F_1(-n_r, 2\gamma|\kappa| + 1; x), \tag{E.1}$$

where $n_r = \epsilon n_a - \gamma|\kappa|$ is, for $u_2(x)$, a positive number or zero, and, for u_1 a positive number (excluding zero).

To determine N_1/N_2 insert these solutions into the second of the equations of (16.23), where a^+ is defined in Eq. (16.29):

$$a^+ u_2 = \left[x\frac{d}{dx} + \left(\frac{x}{2} - \epsilon n_a \right) \right] u_2 = -\epsilon n_a u_1. \tag{E.2}$$

We calculate

$$x\frac{du_2}{dx} = N_2 x^{\gamma|\kappa|} e^{-x/2}$$
$$\left\{ \left(\gamma|\kappa| - \frac{x}{2} \right) {}_1F_1(-n_r, 2\gamma|\kappa| + 1; x) + x\frac{d_1 F_1(-n_r, 2\gamma|\kappa| + 1; x)}{dx} \right\}. \tag{E.3}$$

We now use the recursion relation

$$x\frac{d_1 F_1(-n_r, 2\gamma|\kappa| + 1; x)}{dx}$$
$$= -n_r[{}_1F_1(-n_r + 1, 2\gamma|\kappa| + 1; x) - {}_1F_1(-n_r, 2\gamma|\kappa| + 1; x)]. \tag{E.4}$$

Inserting this in Eq. (E.2), and using Eq. (E.3), yields:

$$N_2 x^{\gamma|\kappa|} e^{-x/2} \left\{ \left(\gamma|\kappa| - \frac{x}{2} \right) {}_1F_1(-n_r, 2\gamma|\kappa| + 1; x) \right.$$
$$- n_r[{}_1F_1(-n_r + 1, 2\gamma|\kappa| + 1; x) - {}_1F_1(-n_r, 2\gamma|\kappa| + 1; x)]$$
$$\left. + \left(\frac{x}{2} - \epsilon n_a \right) {}_1F_1(-n_r, 2\gamma|\kappa| + 1; x) \right\}$$
$$= -(\epsilon n_a) N_1 x^{\gamma|\kappa|} e^{-x/2} {}_1F_1(-n_r + 1, 2\gamma|\kappa| + 1; x). \tag{E.5}$$

The $(x/2)$ terms cancel, and the other ${}_1F_1(-n_r, 2\gamma|\kappa| + 1; x)$ terms have a coefficient $\gamma|\kappa| + n_r - \epsilon n_a$, which vanishes because of (15.47). We are left with

$N_2 n_r = N_1(e n_a)$, or

$$\frac{N_1}{N_2} = \frac{n_r}{e n_a} = \frac{n_r}{\sqrt{n_a^2 - \kappa^2}} = \frac{n_r}{\sqrt{(n_a + \kappa)(n_a - \kappa)}}. \tag{E.6}$$

The normalization condition is

$$\int \left[u_A^2(x) + u_B^2(x) \right] dx = 1. \tag{E.7}$$

Substituting $u_A(x)$ and $u_B(x)$ from Eq. (16.46) leads to

$$\frac{\mathcal{N}'^2}{4}(1 + \epsilon)\left[n_r^2 \int x^{2\gamma|\kappa|} e^{-x} [{}_1F_1(-n_r + 1, 2\gamma|\kappa| + 1; x)]^2 \, dx \right.$$

$$+ (n_a - \kappa)^2 \int x^{2\gamma|\kappa|} e^{-x} [{}_1F_1(-n_r, 2\gamma|\kappa| + 1; x)]^2 \, dx \bigg]$$

$$+ \frac{\mathcal{N}'^2}{4}(1 - \epsilon)\left[n_r^2 \int x^{2\gamma|\kappa|} e^{-x} [{}_1F_1(-n_r + 1, 2\gamma|\kappa| + 1; x)]^2 \, dx \right.$$

$$+ (n_a - \kappa)^2 \int x^{2\gamma|\kappa|} e^{-x} [{}_1F_1(-n_r, 2\gamma|\kappa| + 1; x)]^2 \, dx \bigg]. \tag{E.8}$$

The integrals are calculated similarly to the previous cases:

$$\int x^{2\gamma|\kappa|} e^{-x} [{}_1F_1(-n_r, 2\gamma|\kappa| + 1; x)]^2 \, dx = \frac{[\,\Gamma(2\gamma|\kappa| + 1)\,]^2}{\Gamma(n_r + 2\gamma|\kappa|)}(n_r - 1)!$$

$$\int x^{2\gamma|\kappa|} e^{-x} [{}_1F_1(-n_r + 1, 2\gamma|\kappa| + 1; x)]^2 \, dx = \frac{[\,\Gamma(2\gamma|\kappa| + 1)\,]^2}{\Gamma(n_r + 2\gamma|\kappa| + 1)} n_r! \tag{E.9}$$

so

$$\frac{\mathcal{N}'^2}{2}\left[n_r^2 \frac{[\,\Gamma(2\gamma|\kappa| + 1)\,]^2}{\Gamma(n_r + 2\gamma|\kappa|)}(n_r - 1)! + (n_a - \kappa)^2 \frac{[\,\Gamma(2\gamma|\kappa| + 1)\,]^2}{\Gamma(n_r + 2\gamma|\kappa| + 1)} n_r! \right]$$

$$= \frac{\mathcal{N}'^2 [\,\Gamma(2\gamma|\kappa| + 1)\,]^2 n_r!}{2\Gamma(n_r + 2\gamma|\kappa| + 1)}\left[n_r(n_r + 2\gamma|\kappa|) + (n_a - \kappa)^2 \right] = 1. \tag{E.10}$$

Now use

$$n_r(n_r + 2\gamma|\kappa|) + (n_a - \kappa)^2 = 2n_a(n_a - \kappa). \tag{E.11}$$

This yields

$$1 = \frac{\mathcal{N}'^2}{2} \frac{[\,\Gamma(2\gamma|\kappa|+1)\,]^2 n_r!}{\Gamma(n_r+2\gamma|\kappa|+1)} \, 2n_a(n_a-\kappa),$$

$$\mathcal{N}'^2 = \frac{1}{[\,\Gamma(2\gamma|\kappa|+1)\,]^2} \frac{\Gamma(n_r+2\gamma|\kappa|+1)}{n_a(n_a-\kappa)n_r!}, \qquad \text{(E.12)}$$

$$\mathcal{N}' = \frac{1}{\Gamma(2\gamma|\kappa|+1)} \left[\frac{\Gamma(n_r+2\gamma|\kappa|+1)}{n_a(n_a-\kappa)n_r!}\right]^{1/2}.$$

Solutions to the Exercises

Exercise 3.1 Prove $\epsilon_{ijk}\epsilon_{imn} = \delta_{jm}\delta_{kn} - \delta_{jn}\delta_{km}$.

Solution: We must have $i \neq j \neq k$ if we are to have a non-vanishing contribution. Suppose $i = 1$, then we can have $j = 2, k = 3$ or $j = 3, k = 2$. Also $m = 2, n = 3$ or $m = 3, n = 2$. If $k = n$

$$\epsilon_{123}\epsilon_{123} = \delta_{22}\delta_{33} - \delta_{23}\delta_{32} = 1.$$

If $k \neq n$

$$\epsilon_{123}\epsilon_{132} = \delta_{23}\delta_{32} - \delta_{22}\delta_{33} = -1.$$

We proceed similarly for other permutations of the indices.

Exercise 3.2 Prove $\epsilon_{mij}\epsilon_{nij} = 2\delta_{mn}$.

Solution: If $m \neq n$ we get 0. If $m = n = 1$ then

$$\epsilon_{1ij}\epsilon_{1ij} = \epsilon_{123}\epsilon_{123} + \epsilon_{132}\epsilon_{132} = 2.$$

We proceed similarly for other permutations of the indices.

Exercise 8.1 Prove the identities

$$(a) \quad (\mathbf{p} \times \mathbf{L}) \cdot (\mathbf{p} \times \mathbf{L}) = \mathbf{p}^2 L^2,$$

$$(b) \quad (\mathbf{p} \times \mathbf{L}) \cdot \mathbf{p} = 2i\mathbf{p}^2,$$

$$(c) \quad \mathbf{p} \cdot (\mathbf{p} \times \mathbf{L}) = 0,$$

$$(d) \quad (\mathbf{p} \times \mathbf{L}) \cdot \mathbf{r} = L^2 + 2i(\mathbf{p} \cdot \mathbf{r}),$$

$$(e) \quad \mathbf{r} \cdot (\mathbf{p} \times \mathbf{L}) = L^2.$$

189

Solution:

(a) $\epsilon_{ijk}\epsilon_{imn}p_jL_kp_mL_n = \epsilon_{ijk}\epsilon_{imn}(p_jp_mL_kL_n + p_j[L_k,p_m]L_n)$

$= (\delta_{jm}\delta_{kn} - \delta_{jn}\delta_{km})(p_jp_mL_kL_n + i\epsilon_{kmq}p_jp_qL_n)$

$= \mathbf{p}^2L^2 - p_j(p_mL_m)L_j + i\epsilon_{njq}p_jp_qL_n = \mathbf{p}^2L^2.$

(b) $\epsilon_{ijk}p_jL_kp_i = \epsilon_{ijk}(p_jp_iL_k + p_j[L_k,p_i]) = i\epsilon_{ijk}\epsilon_{kiq}p_jp_q = 2i\mathbf{p}^2.$

(c) $\epsilon_{ijk}p_ip_jL_k = 0.$

(d) $\epsilon_{ijk}p_jL_kx_i = \epsilon_{ijk}p_j(x_iL_k + i\epsilon_{kiq}x_q) = L^2 + 2ip_ix_i.$

(e) $\epsilon_{ijk}x_ip_jL_k = L^2.$

Exercise 8.2 Show that C_1 and C_2 are Casimir operators:

$$[C_1,\mathbf{L}] = [C_1,\mathbf{A}] = [C_2,\mathbf{L}] = [C_2,\mathbf{A}] = 0.$$

Solution:

$$[C_1,L_i] = [\mathbf{L}^2,L_i] + [\mathbf{A}^2,A_i] = A_j[A_j,L_i] + [A_j,L_i]A_j$$

$$= -i\epsilon_{ijk}(A_jA_k + A_kA_j) = 0.$$

$$[C_1,A_i] = [\mathbf{L}^2,A_i] + [\mathbf{A}^2,A_i]$$

$$= L_j[L_j,A_i] + [L_j,A_i]L_j + A_j[A_j,A_i] + [A_j,A_i]A_j$$

$$= -i\epsilon_{ijk}(L_jA_k + A_kL_j + A_jL_k + L_kA_j) = 0.$$

$$[C_2,L_i] = \frac{1}{2}(L_j[A_j,L_i] + [L_j,L_i]A_j + A_j[L_j,L_i] + [A_j,L_i]L_j)$$

$$= -(i/2)\epsilon_{ijk}(L_jA_k + L_kA_j + A_jL_k + A_kL_j) = 0.$$

$$[C_2,A_i] = \frac{1}{2}([\mathbf{L}\cdot\mathbf{A},A_i] + [\mathbf{A}\cdot\mathbf{L},A_i])$$

$$= \frac{1}{2}(L_j[A_j,A_i] + [L_j,A_i]A_j + A_j[L_j,A_i] + [A_i,A_j]L_j)$$

$$= -(i/2)\epsilon_{ijk}(L_jL_k + A_kA_j + A_jA_k + A_kL_j) = 0.$$

Exercise 9.1 Prove that

$$p_r = -\frac{i}{r}\frac{\partial}{\partial r}r$$

is canonically conjugate to r.

Solution:

$$[r, p_r] = rp_r - p_r r = (-i)\left(\frac{\partial}{\partial r}r - \frac{1}{r}\frac{\partial}{\partial}r^2\right) = -i(1-2) = i.$$

Exercise 9.2 Prove that the quantized version of p_r follows from the Weyl–Moyal quantization prescription applied to $\mathbf{p}$ and $\hat{\mathbf{r}}$.

Solution:

$$\frac{1}{2}\left[\frac{\mathbf{r}}{r}\cdot\mathbf{p} + \mathbf{p}\cdot\frac{\mathbf{r}}{r}\right] = \frac{-i}{2}\left[2\frac{\partial}{\partial r} + \left(\frac{\nabla\cdot\mathbf{r}}{r} - \frac{\mathbf{r}\cdot\mathbf{r}}{r^3}\right)\right] = -i\left(\frac{\partial}{\partial r} + \frac{1}{r}\right) = p_r.$$

Exercise 9.3 Prove that p_r is Hermitian.

Solution: We must prove

$$\int[\psi^*(p_r\psi) - (p_r\psi)^*\psi]d^3r = 0.$$

We have

$$p_r\psi = (-i)\frac{1}{r}\frac{\partial}{\partial r}(r\psi),$$

and

$$(p_r\psi)^* = (i)\frac{1}{r}\frac{\partial}{\partial r}(r\psi)^*.$$

Therefore

$$\int[\psi^*(p_r\psi) - (p_r\psi)^*\psi]d^3r$$

$$= -i\int d\cos\theta\, d\phi\left[(r\psi)^*\frac{\partial}{\partial r}(r\psi) + (r\psi)\frac{\partial}{\partial r}(r\psi)^*\right]$$

$$= -i\int d\cos\theta\, d\phi\frac{\partial}{\partial r}[(r\psi)(r\psi)^*]\,dr = -i\int d\cos\theta\, d\phi\frac{\partial}{\partial r}|r\psi|^2dr.$$

For $\lim_{r\to\infty}(r\psi) = 0$ and $\lim_{r\to 0}(r\psi) = 0$ the condition for Hermiticity is satisfied.

Exercise 14.1 Prove $[\mathbf{\Sigma}\cdot\hat{\mathbf{r}}, H] = -\frac{2i}{r}\beta K\gamma_5$.

Solution:

$$[\mathbf{\Sigma}\cdot\hat{\mathbf{r}}, H] = [\mathbf{\Sigma}\cdot\hat{\mathbf{r}}, \boldsymbol{\alpha}\cdot\mathbf{p}] = \Sigma_i\left[\frac{r_i}{r}, \alpha_j p_j\right] + [\Sigma_i, \alpha_j p_j]\frac{r_i}{r}.$$

Calculate the individual terms:

$$\Sigma_i \left[\frac{r_i}{r}, \alpha_j p_j\right] = \Sigma_i \frac{1}{r}[r_i, \alpha_j p_j] + \Sigma_i \left[\frac{1}{r}, \alpha_j p_j\right] r_i$$

$$= \frac{\Sigma_i \alpha_j}{r}[r_i, p_j] + \Sigma_i \alpha_j \left[\frac{1}{r}, p_j\right] r_j = \frac{i}{r}\mathbf{\Sigma}\cdot\boldsymbol{\alpha} - i\mathbf{\Sigma}\cdot\mathbf{r}\frac{\boldsymbol{\alpha}\cdot\mathbf{r}}{r^3}$$

$$= \frac{3i}{r}\gamma_5 - \frac{i}{r}\gamma_5 = \frac{2i}{r}\gamma_5.$$

Also

$$[\Sigma_i, \alpha_j p_j]\frac{r_i}{r} = 2\epsilon_{ijk}\alpha_k p_j r_i \left(\frac{1}{r}\right) = 2i\epsilon_{ijk}\alpha_k(r_i p_j - i\delta_{ij})\left(\frac{1}{r}\right)$$

$$= 2i\alpha_k L_k \left(\frac{1}{r}\right) = \frac{2i}{r}\gamma_5(\mathbf{\Sigma}\cdot\mathbf{L}).$$

We group these terms together and find

$$[\mathbf{\Sigma}\cdot\hat{\mathbf{r}}, H] = \frac{2i}{r}\gamma_5 + \frac{2i}{r}\gamma_5(\mathbf{\Sigma}\cdot\mathbf{L}) = \frac{2i}{r}\gamma_5(\mathbf{\Sigma}\cdot\mathbf{L} + 1) = -\frac{2i}{r}\gamma_5\beta K.$$

Exercise 14.2 Prove $[K\mathbf{\Sigma} \cdot \mathbf{p}, H] = -iK(\mathbf{\Sigma} \cdot \hat{\mathbf{r}})V'(r)$.

Solution:

$$[K(\mathbf{\Sigma}\cdot\mathbf{p}), H] = K[\mathbf{\Sigma}\cdot\mathbf{p}, H] + [K, H](\mathbf{\Sigma}\cdot\mathbf{p}) = K[\mathbf{\Sigma}\cdot\mathbf{p}, \boldsymbol{\alpha}\cdot\mathbf{p} + \beta m + V(r)]$$

$$= K[\mathbf{\Sigma}\cdot\mathbf{p}, \boldsymbol{\alpha}\cdot\mathbf{p}] + mK[\mathbf{\Sigma}\cdot\mathbf{p}, \beta] + K[\mathbf{\Sigma}\cdot\mathbf{p}, V(r)].$$

The individual commutators are

$$[\mathbf{\Sigma}\cdot\mathbf{p}, \boldsymbol{\alpha}\cdot\mathbf{p}] = \gamma_5[\Sigma_i, \Sigma_j]p_i p_j = 2i\epsilon_{ijk}\gamma_5\Sigma_k p_i p_j = 0,$$

$$[\mathbf{\Sigma}\cdot\mathbf{p}, \beta] = 0,$$

$$[\mathbf{\Sigma}\cdot\mathbf{p}, V(r)] = \Sigma_i[p_i, V(r)] = -i\Sigma_i\frac{r_i}{r}V'(r) = -i(\mathbf{\Sigma}\cdot\hat{\mathbf{r}})V'(r).$$

Exercise 14.3 Prove

$$[K\gamma_5 f(r), H] = iK(\mathbf{\Sigma}\cdot\hat{\mathbf{r}})f'(r) + 2mK\gamma_5\beta f(r).$$

Solution:

$$[K\gamma_5 f(r), H]$$
$$= K[\gamma_5 f(r), H] + [K, H]\gamma_5 f(r) = K\gamma_5[f(r), H] + K[\gamma_5, H]f(r)$$
$$= K\gamma_5[f(r), \boldsymbol{\alpha}\cdot\mathbf{p}] + 2mK\gamma_5\beta f(r) = iK\gamma_5\boldsymbol{\alpha}\cdot\hat{\mathbf{r}}f'(r) + 2mK\gamma_5\beta f(r)$$
$$= iK(\boldsymbol{\Sigma}\cdot\hat{\mathbf{r}})f'(r) + 2mK\gamma_5\beta f(r).$$

Exercise 14.4 Prove $K^2(H - \beta m)^2 = K^2(H^2 - m^2 + 2m\beta\frac{Z\alpha}{r})$.

Solution:

$$K^2(H - \beta m)^2 = K^2[H^2 + m^2 - m(H\beta + \beta H)].$$

We have

$$H\beta = m + \gamma_5\beta\boldsymbol{\Sigma}\cdot\mathbf{p} - \beta\frac{Z\alpha}{r},$$

$$\beta H = m + \beta\gamma_5\boldsymbol{\Sigma}\cdot\mathbf{p} - \beta\frac{Z\alpha}{r},$$

$$H\beta + \beta H = 2m - 2\beta\frac{Z\alpha}{r}.$$

Adding these together yields the desired relationship.

Exercise 14.5 Prove

$$(\boldsymbol{\Sigma}\cdot\hat{\mathbf{r}})K\gamma_5(H - \beta m) - (H - \beta m)\gamma_5 K(\boldsymbol{\Sigma}\cdot\hat{\mathbf{r}}) = \frac{-2i\beta K^2}{r}.$$

Solution:

$$(\boldsymbol{\Sigma}\cdot\hat{\mathbf{r}})K\gamma_5(H - \beta m) - (H - \beta m)\gamma_5 K(\boldsymbol{\Sigma}\cdot\hat{\mathbf{r}})$$
$$= K\gamma_5\boldsymbol{\Sigma}\cdot\hat{\mathbf{r}}(H - \beta m) - \gamma_5 K(H - \beta m)\boldsymbol{\Sigma}\cdot\hat{\mathbf{r}}$$
$$= K\gamma_5[\boldsymbol{\Sigma}\cdot\hat{\mathbf{r}}, H] = K\gamma_5(-2i\gamma_5)\beta\frac{K}{r} = \frac{-2i\beta K^2}{r}.$$

Bibliography

Books

[Ada94]   B. G. Adams. *Algebraic Approach to Simple Quantum Systems*. Berlin: Springer-Verlag, 1994.

[BD64]   J. D. Bjorken and S. D. Drell. *Relativistic Quantum Mechanics*. New York: McGraw-Hill, 1964.

[BL81]   L. C. Biedenham and J. D. Louck. *Angular Momentum in Quantum Physics: Theory and Applications*. Reading, MA: Encyclopedia of Mathematics and its Applications 8, Addison-Wesley, 1981.

[Bre91]   D. Bressoud. *Second Year Calculus*. New York: Springer, 1991.

[Bro24]   L. de Broglie. *Licht und Materie*. Hamburg: H. Goverts Verlag, 1930.

[BS57]   H. A. Bethe and E. E. Salpeter. *Quantum Mechanics of One- and Two-Electron Atoms*. Berlin: Springer-Verlag, 1957.

[Dir53]   Paul A. M. Dirac. *The Principles of Quantum Mechanics*. 3rd ed. Oxford: Clarendon Press, 1953.

[DL03]   C. Doran and A. Lasenby. *Geometric Algebra for Physicists*. Cambridge: Cambridge University Press, 2003.

[ER74]   R. Eisberg and R. Resnick. *Quantum Physics of Atoms, Molecules, Solids, Nuclei, and Particles*. New York: J. Wiley, 1974.

[GR65]   I. S. Gradshteyn and I. M. Ryzhik. *Tables of Integrals, Series, and Products*. New York: Academic Press, 1965.

[GS90]   V. Guillemin and S. Sternberg. *Variations on a Theme by Kepler*. Providence, R.I.: The American Mathematical Society, 1990.

[Jac98]   J. D. Jackson. *Classical Electrodynamics*. 3rd ed. New York: J. Wiley, 1998.

[Kel83]   A. Keller. *The Infancy of Atomic Physics: Hercules in his Cradle*. Oxford: Clarendon Press, 1983.

[LL77]   L. D. Landau and E. M. Lipshitz. *Quantum Mechanics*. 3rd ed. Oxford: Pergamon Press, 1977.

[Mer61]   E. Mertzbacher. *Quantum Mechanics*. New York: J. Wiley, 1961.

[MR99]   J. E. Marsden and T. S. Ratiu. *Introduction to Mechanics and Symmetry*. 2nd ed. New York: Springer-Verlag, 1999.

[PGS01]   C. Poole, H. Goldstein and J. Safko. *Classical Mechanics*. New York: J. Wiley, 1998.

[Rig03]   J. S. Rigden. *Hydrogen: The Essential Element*. Cambridge, MA: Harvard University Press, 2003.

[Ros61]   E. M. Rose. *Relativistic Electron Theory*. New York: J. Wiley, 1961.

[Sak67]   J. J. Sakurai. *Advanced Quantum Mechanics*. Reading, MA: Addison-Wesley, 1967.

[SDB04]   W. Siegel, S. Duplin and J. Bagger. *Concise Encyclopedia of Supersymmetry*. Dordrecht: Kluver Academic Publishing, 2004.

[Sea80]   J. B. Seaborn. *Hypergeometric Functions and their Applications*. Berlin: Springer-Verlag, 1980.

[Som24]   A. Sommerfeld. *Atombau und Spektrallinien*. Braunschweig: Friedrich Vieweg Verlag, 1924.

[Sud86]   A. Sudbery. *Quantum Mechanics and the Particles of Nature*. Cambridge: Cambridge University Press, 1986.

[Tak08]   L. A. Takhtayan. *Quantum Mechanics for Mathematicians*. Providence, R.I.: American Mathematical Society, 2008.

[WS10]   O. L. Weaver and D. H. Sattinger. *Lie Groups and Algebras with Applications to Physics, Geometry, and Mechanics*. New York: Springer-Verlag, 2010.

Journals

[AB78]   P. R. Auvil and L. M. Brown. "The Relativistic Hydrogen Atom: A Simple Solution". In: *Am. J. Phys.* 46 (1978), p. 679.

[BB80]   L. Basano and A. Bianchi. "Rutherford's scattering formula via the Runge–Lenz vector". In: *Am. J. Phys.* 48 (1980), p. 400.

[BHJ26]   M. Born, W. Heisenberg and P. Jordan. "Zur Quantenmechanik II". In: *Z. Phys.* 35 (1926), p. 557.

[Bie62]   L. C. Biedenharn. "Remarks on the Relativistic Kepler Problem". In: *Phys. Rev.* 126 (1962), p. 845.

[Bie83]   L. C. Biedenharn. "The 'Sommerfeld Puzzle' Revisited and Resolved". In: *Foundations of Physics.* 13 (1983), p. 13.

[BJ25]   M. Born and P. Jordan. "Zur Quantenmechanik". In: *Z. Phys.* 34 (1925), p. 858.

[BJ53]   W. L. Bade and H. Jehle. "An Introduction to Spinors". In: *Rev. Mod. Phys.* 25 (1953), p. 714.

[Boh13]   N. Bohr. "On the Quantum Theory of Line Spectra". In: *Phil. Mag.* 26 (1913), p. 476.

[Dah95]   J. P. Dahl. "On the Origin of the Runge–Lenz Vector". In: *Int. J. of Quant. Chem.* 53 (1995), p. 161.

[Dar28]   C. G. Darwin. "The Wave Equations of the Electron". In: *Proc. R. Soc. London.* Ser. A118 (1928), p. 654.

[Dir26]   P. A. M. Dirac. "Quantum Mechanics and a Preliminary Investigation of the Hydrogen Atom". In: *Proc. Roy. Soc.* A110 (1926), p. 561.

[Dir28]   P. A. M. Dirac. "The Quantum Theory of the Electron". In: *Proc. Roy. Soc.* A118 (1928), p. 610.

[DJ95]   J. P. Dahl and T. Jørgensen. "On the Dirac–Kepler Problem: The Johnson–Lippmann Operator, Supersymmetry, and Normal-Mode Representations". In: *Int. J. of Quant. Chem.* 53 (1995), p. 161.

[FK05]   M. R. Francis and A. Kosowsky. "The Construction of Spinors in Geometric Algebra". In: *Ann. Phys.* 317 (2005), p. 383.

[GI97]   B. Goodman and S. R. Ignjatoviç. "A Simpler Solution of the Dirac Euation in a Coulomb Potential". In: *Am. J. Phys.* 65 (1997), p. 214.

[Gol75]   H. Goldstein. "The Prehistory of the Runge–Lenz Vector". In: *Am. J. Phys.* 43 (1975), p. 735.

[Gol76]   H. Goldstein. "More on the Prehistory of the Runge–Lenz Vector". In: *Am. J. Phys.* 44 (1976), p. 1123.

[Gor28]   W. Gordon. "Energy Levels of the Hydrogen Atom in the Dirac Quantum Theory of Electrons". In: *Z. Phys.* 48 (1928), p. 11.

[Hei25]   W. Heisenberg. "Ueber quantentheoretische Umdeutung kinematischer und mechanischer Beziehungen". In: *Z. Phys.* 33 (1925), p. 879.

[HI51]   T. E. Hull and L. Infeld. "The Factorization Method". In: *Rev. Mod. Phys.* 23 (1951), p. 21.

[Inf41]   L. Infeld. "On a New Treatment of Some Eigenvalue Problems". In: *Phys. Rev.* 59 (1941), p. 737.

[JL50]   M. H. Johnson and B. A. Lippmann. "Relativistic Kepler Problem". In: *Phys. Rev.* 78 (1950), p. 329.

[JS86]   P. D. Jarvis and G. E. Stedman. "Supersymmetry in Second-Order Relativistic Equations for the Hydrogen Atom". In: *J. Phys. A: Math. Gen.* 19 (1986), p. 1373.

[KCS95]   A. Khare, F. Cooper and U. P. Sukhatma. "Supersymmetry and Quantum Mechanics". In: *Phys. Rep.* 251 (1995), p. 267.

[KK06]   T. T. Khachidze and A. A Khelashvili. "The Hidden Symmetry of the Coulomb Problem in Relativistic Quantum Mechanics: From Pauli to Dirac". In: *Am. J. Phys.* 74 (2006), p. 628.

[Kol66]   M. Kolsrad. "On the Solution of Dirac's Equation with Coulomb Potential". In: *Phys. Noregica.* 2 (1966), p. 43.

[LL74]   J.-M. Lévy-Leblond. "The Pedagogical Role and Epistomological Significance of Group Theory in Quantum Mechanics". In: *Rivista del Nuovo Cimento* 4 (1974), p. 99.

[LR47]   W. E. Lamb and R. C. Retherford. "Fine Structure of the Hydrogen Atom by a Microwave Method". In: *Phys. Rev.* 72 (1947), p. 241.

[MG58]   P. C. Martin and R. F. Glauber. "Relativistic Thory of Radiative Orbital Electron Capture". In: *Phys. Rev.* 109 (1958), p. 1307.

[NKT85]   M. Nieto, V. A. Kostelecky and D. Truax. "Supersymmetry and the Relationship between the Coulomb and the Oscillator Problems in Arbitrary Dimensions". In: *Phys. Rev.* D32 (1985), p. 2627.

[Pau26]   W. Pauli. "Über das Wasserstoffatom vom Standpunkt der Neuen Quantenmechanik". In: *Z. Phys.* 36 (1926), p. 336.

[Pla00]   M. Planck. "Zur Theorie des Gesetzes der Energieverteilung im Normalspektrum". In: *Verhandlungen der Deutschen Physikalischen Gesellschaft.* (1900), p. 237.

[RLB90]   P. K. Roy, A. Lahiri and B. Bagchi. "Supersymmetry in Quantum Mechanics". In: *Int. J. Mod. Phys.* A5 (1990), pp. 1383–1456.

[Rut11]   E. Rutherford. "The Scattering of Alpha and Beta Particles on Matter and the Structure of the Atom". In: *Phil. Mag.* 21 (1911), p. 669.

[SB89]   A. Stahlhofen and K. Bleuler. "An Algebraic Form of the Factorization Method". In: *Nuovo Cim.* 104B (1989), p. 447.

[Sch26] E. Schrödinger. "Über das Wasserstoffatom vom Standpunkt der Neuen Quantenmechanik". In: *Ann. d. Phys.* 75 (1926), p. 361.

[SD91] R. A. Swainson and G. W. F. Drake. "A Unified Treatment of the Non-Relativistic and Relativistic Hydrogen Atom: I. The Wavefunctions". In: *J. Phys. A: Math. Gen.* 24 (1991), p. 79.

[Som16] A. Sommerfeld. "Zur Quantentheorie der Spektrallinien". In: *Ann. d. Phys.* 51 (1916), p. 1 and p. 125.

[Sta98] A. A. Stahlhofen. "Comment on 'A simpler solution of the Dirac equation in a Coulomb potential'". In: *Am. J. Phys.* 66 (1998), p. 636.

[Suk85a] C. V. Sukumar. "Supersymmetry, factorisation of the Schrödinger equation and a Hamiltonian hierarchy". In: *J. Phys. A: Math. Gen.* 18 (1985), p. L57.

[Suk85b] C. V. Sukumar. "Supersymmetry and the Dirac equation for a central Coulomb field". *J. Phys. A: Math. Gen.* 18 (1985), p. L697.

[TT93] D. Tangerman and J. A. Tjon. "Exact Supersymmetry of the Non-Relativistic Hydrogen Problem". In: *Phys. Rev.* A48 (1993), p. 1089.

[Wal79] S. Waldenstrøm. "On the Dirac Equation or the Hydrogen Atom". In: *Am. J. Phys.* 47 (1979), p. 1098.

[Wil15] W. Wilson. "The Quantum-theory of radiation and Line-spectra". In: *Phil. Mag.* 29 (1915), p. 795.

[Wit81] E. Witten. "A new proof of the positive energy Theorem". In: *Commun. Math. Phys.* 80 (1981), p. 381.

[Wit90] E. Witten. "Spontaneous Breaking of Supersymmetry in Quantum Mechanics". In: *Int. J. Mod. Phys.* A5 (1990), pp. 1383–1456.

[Won86] M. F. K. Wong. "Infeld–Hull Factorization and the Simplified Form of the Dirac–Coulomb Equation". In: *Phys. Rev.* 34 (1986), p. 1559.

[WY82] M. F. K. Wong and H.-Y. Yeh. "Simplified Solution of the Dirac Equation with a Coulomb Potential". In: *Phys. Rev.* D25 (1982), p. 3396.

[WZ74] J. Wess and B. Zumino. "Supergauge Transformations in Four Dimensions". In: *Nuclear Phys.* B70 (1974), p. 39.

Index